R. J. BROPHY
111 S. SIXTH ST.
MT. HOREB, WIS. 53572

2-15-89

the
Electric
Windmill

AN INADVERTENT
AUTOBIOGRAPHY

the
Electric
Windmill

AN INADVERTENT AUTOBIOGRAPHY

by
TOM BETHELL

REGNERY GATEWAY · Washington, D.C.

"Agnostic Evolutionists" by Tom Bethell reprinted courtesy of *Harper's Magazine*.
Copyright © 1985 by *Harper's Magazine*. All rights reserved. Reprinted from the
February issue by special permission.

"Darwin's Mistake" by Tom Bethell reprinted courtesy of *Harper's Magazine*. Copyright © 1976 by *Harper's Magazine*. All rights reserved. Reprinted from the February
issue by special permission.

Library of Congress Cataloging-in-Publication Data

Bethell, Tom.
    The electric windmill / by Tom Bethell.
        p.   cm.
    ISBN 0-89526-568-0: $17.95
        1. United States—Civilization—1945–   2. United States—Popular
    culture—History—20th century.   3. United States—Politics and
    government—1945–   I. Title.
    E169.12.B43   1987
    973.9—dc 19
                                                                87-35441
                                                                CIP

Published in the United States by
Regnery Gateway
1130 17th Street, NW
Washington, D.C. 20036

Distributed to the trade by
Kampmann & Company, Inc.
9 E 40th Street
New York, N.Y. 10016

10   9   8   7   6   5   4   3   2   1

For all the Fabillis

Contents

the Electric Windmill

AN INADVERTENT AUTOBIOGRAPHY

Introduction

Most of the articles in this book were earlier published in various magazines, frequently in different form. I have rewritten many of them.
When you look back at something you wrote ten years earlier, the rough
patches and the deadwood stand out clearly. So I have done a lot of
pruning and revising. The only exceptions are the two articles on evolution reprinted from *Harper's* magazine. These were subsequently
debated (and attacked) in the academic and popular scientific press, and
so it seemed best to leave them as they were rather than touch them up.

In a good many of the articles I frequently appear as an observer off to
one side, notepad or sometimes tape recorder in hand. Taken together
the pieces therefore constitute an inadvertent autobiography, one that
could have been called "My Life in America." This was not what I had in
mind when I embarked on the arduous and time-consuming task of
putting this collection together, but it does create a theme.

It did strike me as a good idea to exclude the numerous policy-oriented
articles that I have written over the years. Most of the pieces here
describe some scene or person, written more for entertainment than
instruction. Some textbookish stuff still seems to have slipped through.
Like tropical vegetation, it keeps growing back. It is hard to eradicate it
entirely.

Apart from a handful of articles written for the (London) *Financial
Times* between 1969 and 1971, the first reporting I did was for the New
Orleans *Vieux Carre Courier*, starting in 1972. This was an eye-opening
experience for me, almost literally. Before then I had assumed that
writers have "interests," about which they store up knowledge and ideas,
like capital. Then they "spend" these savings in the form of books or
articles about their fields of specialization. It followed that there were
only a limited number of things one could write about. One would have

to be careful not to splurge when putting pen to paper! But with journalism I made the discovery that the preliminary accumulation is not really necessary. I found out (no one had told me!) that it was possible to visit some scene and write down what people said, what they were wearing and what they looked like without even being particularly interested in the subject at hand. In fact, in my case, such "disinterestedness" actually seemed to benefit the writing.

My first and perhaps best editor, Jeannette Gottlieb Hardy at the *Vieux Carre Courier*, soon brought this home to me. She would be polite when I brought her my earnest pensees about music (precious offerings from my intellectual savings account, as I thought), but I could see she wasn't too interested. But she would be delighted by what I thought of as ephemeral piecework—in one ear and out through the pen the same day—about some local rascal seeking office. After a while I learned to trust her judgment in such matters.

I started to write for the overriding reason that I needed the money, and since then I have often thought about the strange, on-the-job literary education I received in New Orleans. The journalist turns his own unfamiliarity with a subject to advantage. He is able to transmit to the reader the pleasure of discovery, of understanding something for the first time, in a way that is nearly impossible for someone who has been studying a subject for years. How often have we met the expert who cannot explain his field because he can't recall the mental state of *not* understanding it?

The first two chapters are new. For years I accumulated notes on William Russell but found it impossible to write anything, I think for the very reason that I was too interested in his pursuits. A 12-year hiatus has, I hope, given me enough detachment. Chapters three, four, five and seven were originally published in the *Vieux Carre Courier* in 1973, although they have been revised. Chapters six and eight came out in *The Washington Monthly*, in 1975 and 1976, again in different form. Probably half the revised "Sirhan Sirhan" is new.

The long, segmented chapter on Los Angeles, "Beyond the Ochre and the Umber," originally appeared as columns in *The Los Angeles Herald Examiner* (I have included here about two-thirds of the columns I wrote for that paper over a two-month period in 1981). The two pieces "Porgy and Bass" and "Critic as Building Inspector" originally appeared in *Chronicles of Culture* in the late '70s, but I have rewritten them. "Sonia Johnson's Constitutional," again in different form, started life in *Chief Executive* magazine.

Everything else was first published in *The American Spectator*, for which I have had the pleasure of writing a monthly column since 1977. Some of these pieces were fairly light-hearted and not meant to be taken too seriously. Oddly enough, they seem to have stood the test of time better than other more solemn compositions that I slaved over for weeks.

I would like to take this opportunity to thank a number of people whose help has been indispensable and who have been very much a part of my life in America. When I first showed up at the *Vieux Carre Courier*, the paper was just then blossoming into a lively weekly under the editorship of Jeannette Hardy. Her husband had recently bought the paper and she had become editor more or less by default, I think, with no previous experience. But she had enough confidence in her own judgment to believe that if something interested her it would also interest the readers; that was her great strength as an editor (along with a willingness to work 12-hour days). Anyway, thanks, Ginny. Without you I'm not sure I would ever have got started in journalism.

There's nothing better than working for a weekly, and the next 18 months or so at the *Courier* I look back on with great pleasure. James K. Glassman had just arrived in town from Boston or thereabouts, and with him Jack Davis, to start another weekly paper, the *Figaro*. Charlotte Hays from Greenville, Miss., was also at the *Courier*, and one day a high school senior named Nick Lemann came into the office with an amazing, *Esquire*-style "Guide to Local Prep Schools." Today he is national correspondent for the *Atlantic Monthly*. We all had great fun competing with one another and trying to show off the latest journalistic techniques; on Wednesday night at about 11 p.m., one of us would make a quick sortie to pick up the latest *Figaro*, fresh off the press. Generally we had 'em cold. . . .

Alas, these great moments don't last and by 1974 I moved on to *New Orleans* magazine with Jeannette Hardy. (She and her husband had sold the *Courier* to Philip Carter). But working on a city monthly just wasn't the same, somehow, and by 1975 I accepted Charlie Peters' kind offer to become an editor at *The Washington Monthly*. A little over a year later Lewis Lapham, the editor of *Harper's*, asked me to become a Washington editor of that venerable journal. Charlie and Lewis could hardly have been more different as editors, yet I learned a great deal from both of them and am grateful for all they did for me.

I first saw *The American Spectator* (then called *The Alternative*) in the New Orleans Public Library. I was overjoyed—at least there was one magazine that didn't take itself too seriously (to my mind the besetting sin of American journalism). I think I knew it wouldn't be very long before I wrote for it. I have now been doing so for ten years, with great pleasure. My thanks to Bob Tyrrell, who has given me great freedom, never telling me what to write or how to write it; and to his succession of unfailingly cheerful managing editors—Adam Meyerson, Steve Munson, Eric Eichman and Wladyslaw Pleszczynski.

I am grateful also to William F. Buckley, Jr., who has generously welcomed me into the pages of *National Review*. My work for that magazine, dealing mostly with the elements of economics, is not included here, but I hope that its essence will be contained in my next book.

My thanks also to Mary Anne Dolan, who invited me to write a weekly article about Washington social life for *The Washington Star* when that paper was still edited by Jim Bellows. I think I subconsciously heaved a sigh of relief when the *Star* was sold four weeks later. Reporting on social events in Washington turned out to be trickier than I had imagined. Bland coverage bored the readers; spicy reporting offended the people you were writing about. I could see no solution to this dilemma, but was saved by the investment bankers. Mary Anne Dolan then moved on to *The Los Angeles Herald Examiner* with Bellows and in 1981 she invited me to fill the "writer in residence" position there for two months. I much enjoyed it and learned a lot.

A special thank you to three people. George Gilder I met in 1978, and he and his wife Nini have become good friends, welcoming me into their family and honoring me as godfather to their daughter, Mellie. I have greatly valued George's persistent and always good-humored encouragement, not to mention his insights. Joe Sobran and I spoke at the same conference in June 1980, and since then he, too, has become a good friend. We must have spent hundreds of hours on the phone together, many of them trying to unravel the mysteries of the Hive. Joe's originality of mind has been a constant source of inspiration to me. Through thick and thin since our *Courier* days together, Charlotte Hays has remained a close friend. Her literary talent will I believe soon reach the wider audience it deserves.

Finally, my thanks to W. Glenn Campbell, the director of the Hoover Institution, where this book was compiled and rewritten—usually late at night. There can be no greater privilege for a writer than a stay at the Hoover Institution, and no more rewarding place to work.

July, 1987

I Was a
Wishy-Washy
Liberal

I arrived First Class on the Queen Elizabeth. Characteristically, I had postponed calling Cunard in London for a reservation until it was almost too late. By then First Class was all they had. So I sailed past the Statue of Liberty in a style that would have been unfamiliar to the huddled masses, as it was to me. Not that I thought of myself as an immigrant. Or did I, I now wonder? Certainly I brought along rather a lot of baggage, as though I knew I would stay. Looking back now after so many years it's hard to say with certainty what I had in mind. I wasn't thinking in terms of "career," I know that. A few months earlier I had graduated from Oxford University, and like many of my friends there I wanted to see something of America. Above all, I wanted to go to New Orleans, whose status as the birthplace of jazz made it my Mecca. I had a job teaching at a prep school in Virginia, and I imagined I would stay in the country for a year, maybe two at the outside. The date of my arrival was August 28, 1962.

Looking back now, I am conscious that my total ignorance of America was exceeded only by the confidence with which I criticized it. From the moment I stepped off the ship I was opinionated about all that I saw. Not that my opinions were really my own. They had been unconsciously absorbed while growing up in England. It embarrasses me now to recall the condescending litany of educated British opinion about Americans, who were said to be: childlike, rich as a rule, sometimes quite amusing, often generous, uncultured of course, and often possessed of a surprising degree of expertise in technical topics such as engineering or whatever:

boring, but for all one knew, important. Americans were said to be good at tinkering with gadgets. In addition, it was understood that they were themselves frequently aware of their shortcomings and more than willing to take self-improvement lessons from educated Englishmen such as myself.

As the Queen Elizabeth was nudged by tugs into the Hudson River dock I gazed out at the stately skyline dominated by the Empire State Building. Secretly I was impressed, but I believe I affected a supercilious air. A crew-cut, Madras-jacketed American at my elbow remarked that I would not have to worry about Red Indians shooting arrows in my direction. Nor were the skyscrapers in imminent danger of falling down: a wiseguy, as the Yanks would say. I laughed and thanked the man for his advice. But was I not the missionary in heathen territory? Should I not already be dispensing advice rather than receiving it?

What a different time that was! Kennedy was president, but in some ways the 'fifties were still in progress. The men had crew cuts, narrow ties and serviceable, washable, straight-line suits seemingly a size too big (not "built," or contoured in the Savile Row style). New York was on top of the world at that time, as I can now see. It was not considered unsafe to go to Harlem, or indeed anywhere. Dennis Cherlin, an American I had known at Oxford, was kind enough to meet me at the Cunard terminal, and a couple of days later he took me to a Broadway play, *A Thousand Clowns*, with Jason Robards. I can remember nothing about it except that I enjoyed it. The city was sunny and tremendously hot. Steam seeped mysteriously out of manhole covers in the middle of the lumpy streets. The New York subway was safe and innocent of graffiti, but the subway system itself was baffling to a Brit. (Not that anyone then used the word Brit, which seems to have emerged only in the 1970s.)

That first Labor Day Weekend I went by train from Grand Central Station to Greenwich, Connecticut, where I met Hamish McDougall. He had already taught for a year at Woodberry Forest School in Virginia, which is where I was headed. If East Coast prep schools are Anglophile today, they were even more so then. In the summer of '62 McDougall had been tutoring a boy called Dupont in Greenwich. That had led to an invitation to spend Labor Day Weekend in Prout's Neck, Maine. When I telephoned Hamish to ask him about the ride down to Virginia he told me it would be okay to join him for the weekend in Maine. So we drove up together from Greenwich. (Poor Hamish was killed less than a year later in a car crash in Virginia's Blue Ridge Mountains.)

It was an upper-crust gathering at Prout's Neck. Socially, I like to tell people, it has been downhill for me in America ever since. (Perhaps I should make an exception of the weekend I spent at the Rockefeller estate, Pocantico Hills, described later. There I met Happy Rockefeller by the swimming pool and Henry Kissinger inside it.) I can't remember the names of the half-dozen or so people at the Maine house party, except for a young lady named Marnie Linen. Her father was a big wheel at *Life*, I was told, and fortunately I was told this before I made any disparaging comments about the magazine, a copy of which was to be found on the breakast table as I went down the first morning. Marnie was at the table, too.

I remember in particular that there was a young man who, when asked what university he had attended, replied "Harvard and Yale," but managed to affect a self-deprecating air as he did so. Everyone was very nice. We played charades in the evening. (There, I told you Americans were childlike.) We played tennis in the morning. One evening we went to a cliffside lobster and corn-on-the-cob cookout: in retrospect, Kennedyesque . . . I recall that weekend now as an incredibly remote, inaccessible and unrealistic event—Camelot! Ever since Labor Day Weekend has for me had a flavor of nostalgia. Perhaps it is the same for all Americans, with its last look at the vanishing summer.

Even though I had been in the country for less than a week, I didn't hesitate to suggest various ways in which national customs and folkways could be improved. Mercifully I have now forgotten what I said. But looking back, what surprises me as much as my own arrogance is my hosts' tolerance. Indeed, tolerance is too weak a word. They actively encouraged my naive comments, sought them out and gave them a consideration they certainly did not merit. I even remember thinking at the time that no American newcomer to England would dream of criticizing the British in analogous surroundings. Nor would the British dare to speak in this way to the French, say, or indeed to anyone excepting Americans.

Tom Wolfe has written bitingly about penniless Brits in New York who are rudely condescending to the Americans they sponge off. I know only too well what he means! But in defense of the British I will add that they would be less likely to behave in this way if the Americans they met were less thirsty for accusation. Now that I am an American myself—I was naturalized in New Orleans in 1974—I can say without once again playing the insolent guest that this American penchant for self-criticism seems to have grown stronger in the past quarter-century. It is a most disturbing and destructive trait—even if it is mostly confined to America's hereditary (dare I say anti-American?) elites. (At bottom they seem to resent the egalitarian, anyone-can-play structure of U.S. society.) In any event, it is a testament to the strength and greatness of the country that it has managed to absorb so much hostility from its most privileged citizens.

Like most Britons in America I came to my senses after a while and realized what a lot of nonsense I had been permitted to get away with. I think I found that this took about six months from the moment of stepping off the boat. (Others I know have experienced a similar time lag.) Of course there are notorious exceptions—anti-American Brits who find symbiotic companions in Manhattan and try to avoid venturing west of the Hudson River or south of the Statue of Liberty. But after the passage of a few months most British immigrants are to be found pensively staring down at the sidewalk, for all the world as though they are trying to remember where they have left some personal item. In fact, they are recalling with embarrassment some vile rudeness and vowing not to let it happen again.

I want to reconstruct my political outlook at that time, not because I had given politics any thought, but precisely because I had not. In this I am sure I was typical of most university graduates. Such matters as government intervention in the economy I had never even considered. Nonetheless, if pressed, I would have described myself as a liberal, and in conversation I would have sounded like one. The point is that, although I knew nothing about politics or government or economics, I did know which political responses were considered in good taste and which were offensive to polite society: which evoked a murmur of assent and which elicited looks of worried reappraisal.

And so, by responding to such social cues, one's politics were acquired as a form of manners. In this way one vaguely joined the progressive camp, without so much as giving the substance of policy a minute's thought. The appropriate political response was learned in the same way that one learns the unwritten rules of etiquette. But then unwritten rules are harder to challenge than written ones.

Let me give an example of the way this worked. About a decade before I arrived in America there had been something called McCarthyism. I knew nothing about it, beyond understanding that there had been this . . . rather *gauche* man (not exactly *le mot juste*, come to think of it!), a U.S. senator apparently, who had blotted his copybook in this very public, very nasty way, by going around calling perfectly innocent people Communists, thereby ruining their lives.

So if someone mentioned "McCarthy" or "McCarthyism" one knew how to respond, just as one would know—let us hope!—which fork to use at dinner. The correct response would be: a little shudder of distaste and perhaps some such comment as: "Frightening to think of really, isn't it?" That would be very satisfactory and the assembled company would find no fault. On the strength of such observations one would satisfy political etiquette in Manhattan, in Greenwich, Connecticut, and in Prout's Neck, Maine. No need, thank God, actually to read up about Cohn and Schine, or whoever, or memorize details about the Underwood typewriter or the Prothonotary Warbler. Certainly one didn't have to be ready to provide the names of innocent non-Communists whose careers had been blighted by McCarthy's recklessness. Which was a mercy and a blessing, because it would be pretty bloody boring to have to read up on all that stuff, wouldn't it? All you had to do was assume the worst about McCarthy.

In such etiquette-dominated discourse, genuinely held political or philosophical convictions tend to remain hidden. Propriety overrides principle. But above all, as I say, it is not necessary to have any such convictions. It is necessary only to make the socially correct response. This has been a key feature of liberalism at least since I have been in the United States. It is an ideology that always remains inexplicit and undefined, devoted ostensibly to such procedural considerations as "freedom of speech" or "equal treatment under the law," and experienced only at the level of etiquette by most of those who assent to it. In this it is analogous to table manners. We do not normally write down such unwritten "rules" of dining, or make explicit the social goal that they are intended to achieve. We merely (as a rule) obey them.

Wishy-washy liberalism is as good a name as any to describe a philosophy that recruits its adherents through implicit social intimidation. Such was my political position when I first came to America, as it has also been the position of many university graduates before and since. (Curiously enough, the far more explicit leftism of today's professoriat is likely to raise the hitherto undissected liberal ideology to the level of consciousness, risking the explicit rejection of it.)

I was a wishy-washy liberal, then. Did I know anything about McCarthyism when I came to the U.S.? No. Did I sympathise with Communism in the slightest? No. Did I actually believe there had been a brief reign of terror for progressive intellectuals in the 1950s? No-o. Did I respond with appropriate nose-curling when the subject of McCarthyism came up? Of course.

From Greenwich I drove south with Hamish, across the George Washington Bridge, down the New Jersey Turnpike (then recently completed), over the Delaware River, and then by back roads, tobacco roads really, across Maryland, as there was no interstate highway between New York and Washington then. We drove through Washington in the dusk, and by the time we arrived at Woodberry Forest, 80 miles south-west of Washington, it was long since dark.

In those days Englishmen were more in demand at private schools, as in America generally, than they are today. Out of several teaching offers I chose the one in Virginia because I knew it to be in the south and hence imagined it to be near New Orleans. But I couldn't help noticing that it took all day to drive from Greewich, Connecticut, to Orange, Virginia—despite the relative propinquity of these two spots on a map of the United States! By nightfall my suspicions were aroused. I was dismayed to find that New Orleans was a further thousand miles down the road. It would hardly be feasible for me to go there at weekends. As it turned out, I wouldn't have time anyway.

On my first night in Virginia an invisible string section of cicadas was tuning up, and then sawing and sighing endlessly. As I looked out at the dark, hot night from my strange room in Turner Hall, I felt lonely for the first time in the New World.

The next day, in ninety-degree heat, a crew-cut football squad was shoving padded barricades across a stubbly field. It was beautiful country. On the horizon a range of cool blue mountains stretched from north to south as far as the eye could see. The main school building had a tall, post-bellum portico, with a shiny Buick or two in front, shimmering in the heat. Inside, a courtly, stout Negro with polished head was polishing the floor and exchanging courtesies with everyone who came and went. Had I not seen him in *Gone with the Wind*?

One evening the headmaster gave a reception at "the residence," an 18th century structure. Cookies, ice-cream and soda-pop were served for the faculty, as though it were a children's tea party. But this was perfectly normal, apparently. The wives seemed to give little curtsies as they came into the room. Everyone spoke with elaborate courtesy. Today I have the feeling that this old south has indeed gone with the wind, but it was still lingering on in quiet pockets of the Virginia Piedmont in the early 1960s.

The students (boys aged 14 to 18) turned out to be much more lively and intelligent than I had expected. Almost all were from southern states. For the most part their fathers were businessmen. The boys arrived at the school without the precocious, inky, Latin learning instilled by cane at

English preparatory schools, but they seemed to catch up quickly and by graduation were as well prepared as their counterparts at the better English public (that is, private) schools.

My impression was that at grade school the boys had studied Creative Wisecracking and little else. I was endlessly amused by their repartee. I was supposed to be teaching them mathematics, about which I was slightly better informed than they. But there were many diversions and I was always happy to let quadratic equations take a back seat to their accounts of disreputable episodes back home. I can see them now as I write, sprawled in their desks in their Madras jackets, white socks and loafers, with their shaven heads and laconic, show-me expressions—forever looking for an opportunity to debate and argue.

"Mister Bethell, would you rather be red than dead?"

I would wriggle out of that one by saying I would rather be neither.

They were conservative, to the extent that teenaged boys can be said to have political opinions. They loyally repeated what they had heard their parents say. Since I was from England, people at the school assumed, without too much inquiry, that I supported the progressive position on such matters as the Cold War and "socialized medicine." (Why do we not hear that phrase any more by the way?) Their assumption was correct. I too would have regarded it as a betrayal of my background to have forsaken the polite progressivism that identified one as a man of taste and discernment.

One day the headmaster, a remarkable Texan named A. Baker Duncan, prognathous, bald, immensely tall and immensely knowledgeable about everything that was going on in the school, asked me to speak to the "stoodent bahdy" about the Cuban missile crisis, then very much in the news. I recall standing up on the stage and quoting from a James Reston column and a (London) *Observer* editorial, and perhaps also something from *The New Statesman*, although what arguments I used I have now forgotten. A few of the seniors gave me a hard time, attacking me from somewhere to the right of Barry Goldwater. Baker Duncan told me later that he was pleased because I had "made them think." Certainly they had made me think.

Within two or three years all those students became a part of the "sixties generation" in college. I think it would have been impossible, in the fall of 1962, to have predicted this development. What we now mean by "the sixties" did not begin until the mid-sixties—I would put it at 1965. Years later when I told Jude Wanniski, the author of *The Way the World Works*, that I had come to America in 1962 he said with uncharacteristic pessimism that I had at least had three good years. I knew what he meant.

In 1974, in New Orleans, I saw one of the senior students who in 1962 had so hawkishly supported the removal of Soviet missiles from Cuba and had scorned my wishy-washy views. He was an intelligent young man, always had good grades, and had gone on to a good university in the South. Now he was sitting on a mat in the open-air flea-market a block from the Mississippi River. He had a glazed look in his eyes and his hair was down to his waist. A few pitiful objects were set in front of him, apparently for sale. I do not know if he had gone to Vietnam—I think not. He did not see me and I stayed out of his line of vision. By then I was beginning to read occasional issues of *National Review* and *The American Spectator* in the New Orleans Public Library.

It was art, not politics, that had attracted me to America, and it took me a long time to figure out that there was a connection between the two. During my first decade in America I accepted it as a given (so deferential, you see, to political etiquette) that America was politically backward; that is, insufficiently like Western Europe. But even before I stepped on board the Queen Elizabeth I was well aware that America was culturally advanced; that is, mercifully unlike Western Europe.

The one subject that I knew something about by the time I left England was music. Ever since I had listened as a child to a 78 rpm recording of Bach's Suite No. 3, and then to a Mozart piano concerto (K.488), I had always found it easy to "understand" music, and up to a point my teenaged musical taste conformed closely to the conventional judgment. Bach, Mozart and Beethoven I loved. (An effortless bull's-eye!) Beyond that, I broke completely with the consensus. It seemed clear, at least to my ear, that with the romantics things began to unravel very quickly. And by the early 20th century it was surely indisputable that the Western classical tradition had not only fallen apart but its residue was under deliberate attack by a brazenly destructive avant-garde.

The only problem was that no one seemed to be saying what was to me so obviously true! There was this polite pretense that everything was still wonderful in the realm of music—different, of course, but just as good in its own way. Hadn't one heard? Composers never were appreciated in their own day. And so on *ad nauseam*. I didn't believe a word of it. But my dissent was lonely. Then one day, while I was at Oxford, I came across a book called *Death of a Music?* by an American named Henry Pleasants. I read it excitedly, and then reviewed it painstakingly—with much rewriting and discovering how difficult it was to say what you meant, and finding out at the same time that you didn't know *what* you meant until you said it; all of which made writing very nearly a logical impossibility. But eventually the review appeared in *Isis*, an undergraduate magazine. It was my first time in print.

I sent the review to Mr. Pleasants, who turned out to be a diplomat stationed at the American Embassy in Bonn. He soon replied and asked me to come and see him on a particular evening because he would be in London, staying with the Bruces; his wife was giving a harpsichord recital at the Wigmore Hall. Pleasants himself had been a music critic with one of the Philadelphia newspapers. We did meet and I felt proud to shake his hand. He even asked me if he could possibly write an article for *Isis* replying to his numerous critics. His book had been fiercely attacked by reviewers, almost all of whom were slavishly in the avant-garde camp, for reasons that I could not then understand (and cannot to this day).

Years later, in all the furious revelations of Watergate, it turned out that the amiable Mr. Pleasants had all along been employed by the Central Intelligence Agency. How the modernists must have felt vindicated! The irritating critic crying out from the sidelines that the emperor had no clothes was, after all, a CIA agent! And so could be expected to mislead us, in art as in politics. But not in my book. He was right then and he is right today.

Pleasants made the point that just as European music had declined, a new and more vital American idiom had arisen to replace it. The new music was both popular and yet sufficiently complex to appeal to those with discriminating taste. Moreover, its appeal was worldwide. But he didn't need to tell me any of this because I was by then a regular at Dobell's Jazz Record Shop on Charing Cross Road. I was forever on the lookout for copies of William Russell's rare American Music label (10-inch LPs pressed in minute editions on transparent red vinylite). Russell's recordings of Bunk Johnson, George Lewis and others, made in New Orleans dance halls in the 1940s—the so-called "New Orleans revival"— had enjoyed an underground reputation in the England that I grew up in. By the time I went to Oxford in 1959 I was spending about half my waking life listening to New Orleans jazz records. Later I played a very amateurish trumpet in a jazz band at the university. It never ceased to amaze me that from time to time we were hired and paid real money to play, sometimes on river barges, sometimes even for debutante dances, although in that case always in a subordinate, non-starring role.

After arriving in America it took me a long time to realize that the things I approved of culturally were a natural expression of the things I disapproved of politically. The liberal wish politically (by which I mean, of course, the illiberal wish) is for many of the people's decisions to be made for them by their betters in Washington, armed with the power of coercion, and employed at the people's expense. I had not seen, and it took me at least ten years to see, that the popular American culture that I thought original and creative had emerged from the American political culture that I thought unsophisticated.

Liberals who like jazz are in the confused position of approving both elitist policy and populist culture—a point that has not been sufficiently grasped by the Marxist historian Eric Hobsbawm, who wrote jazz criticism for *The New Statesman* (under the name Francis Newton), and whose columns I used to read in those days. But at that time I suppose Marxists could still cherish the illusion that their goal really was to restore power to the people. One day Hobsbawm gave a talk on jazz to a very small group at Oxford, and I went to hear him. In those days politics were of no interest to me, and I was unaware of Hobsbawm's political bias. The one thing I can recall was his aversion to Henry Pleasants.

In December, 1962, I arrived in New Orleans by Trailways bus from Charlottesville, Virginia, a journey of almost two full days that I vowed never to repeat. But my goodness it was exciting to be there, with the street cars clanging so familiarly past the palm trees on Canal Street. It is amazing what a little background knowledge, acquired from youthful reading, will do to transform the humdrum into the romantic, the mundane street into the old acquaintance, the childhood friend recognized with delight. It was something to be on Burgundy Street—with George Lewis somewhere in the city at that very moment. Can you imagine how a Beethoven-lover might feel to be transported to the Vienna of 1820, there to make inquiries, and to be told, "Old Ludwig? The one that's hard of hearing? In the evening he's generally to be found at that coffee shop down the street there . . . ."

That's how I felt on my first day in New Orleans.

Eighteen months earlier Preservation Hall had opened on St. Peter Street, in the heart of the French Quarter. This would become a well-known forum and platform for the older New Orleans jazzmen who played in the traditional or "Dixieland" style. (Among the traditional jazz enthusiasts in England, the word Dixieland denotes a slicker style of playing, epitomized by Al Hurt and Pete Fountain. European devotees of

New Orleans music are always bothered when Americans nod in agreement and say that they too like "Dixieland," thereby arousing the suspicion that the distinction between the artistic and the commercial becomes more and more difficult to discern the closer you get to your own backyard.)

Anyway, on my first afternoon in the city I went to Preservation Hall. The iron gate was open and inside the dusty, run-down hall, with a damp, mildewy flavor all its own, there sat the old-time trumpeter Punch Miller, with rather a hang-dog look on his face. "Come on in," he said to me. Had I brought along a trumpet, he inquired? He was just about to give a lesson and any number could join in. Also in the hall was a high school student with a clarinet—Tommy Sancton (now with *Time*). I said I had left my trumpet in England—didn't know I'd be needing it. "Well that's all right," said Punch, not put out at all. "We can find you one here."

That evening you could hear the ships' horns on the Mississippi River, two blocks away, and the calliope on the riverboat hooted out a hilarious *Darktown Strutters' Ball* half way across the city. Rare are the moments that are long anticipated and yet do not disappoint, but this was one of them. George Lewis, "Kid" Howard, Jim Robinson, Joe Watkins and "Slow Drag" Pavageau played at Preservation Hall that night. I found it hard to believe that the audience was so small.

There was one disappointment. William Russell, who had made the wonderful recordings of Bunk Johnson, George Lewis and others in the 1940s, was not in the city. He had left for an indefinite period to take care of his aging parents in Canton, Missouri. I wanted to talk to Russell almost more than I wanted to hear the remaining musicians.

William Russell Wagner

In June 1963, I drove West for two days from Charlottesville, Virginia, in my baking hot Peugeot. I found myself in Mark Twain country. Bill Russell lived in a white frame house in Canton, Missouri, a few short blocks from the Mississippi River. His father had built the house and still lived in it, and Bill Russell himself had been born in it in 1905.

In a way, I had come to America because of Bill Russell. He had made some World War II jazz recordings in New Orleans that had strongly attracted me in England. He had issued these rarities himself on his own label, American Music, in tiny editions of a few hundred copies. The liner notes on his record sleeves were eccentric—with various barbs, pacifist comments and a sense of disenchantment with the American Way unusual among record producers, or indeed Americans (as I then thought). He was said to have kept many unissued master recordings. I was anxious to here them, and meet him.

A professorial-looking gentleman came to the door and peered out, holding a violin. He was wearing a loose white shirt with sleeves rolled up, spectacles, and was bald with a fringe of white hair in back.

"Are you Mr. Russell?"

"Oh, yes."

I had written, telling him I wanted to see him, and he had written back saying it would be okay to come. But he was not too encouraging, reminding me of his invalid parents.

He sat me down in a kind of swing on the front porch and returned with the manuscript of a book about Warren "Baby" Dodds, the New Orleans drummer who went to Chicago with King Oliver and Louis Armstrong. Later I met half a dozen other people who told me that Bill Russell had used his Dodds manuscript as a getting-to-know-you device with them.

I stayed in Quincy, Illinois, about fifteen miles south on the other side of the river, and returned to see Bill every day for five days. I think it was on the second day that he allowed me into the house. He lived in the ground floor front room, where there was a bed, piles of cardboard boxes, violin paraphernalia, a parakeet and a birdcage. The bird was called Pretty Baby, named after a tune by "Professor" Tony Jackson, a Storyville pianist. Russell spent a good deal of those five days telling me that his bird was more intelligent than human beings—of whom he had a low opinion, he said.

Bill Russell's parents were both in their eighties. I never did meet them, although once I heard Bill's father call out "Russell!" commandingly from a back room. And I caught a glimpse of his mother, haggard in a wheel chair. One morning was spent repairing the lawn mower and cutting the grass, another time Bill told me that he had a roast in the oven. He was cooking all the meals. Russell had never married—"never had time," he told me several years later in New Orleans.

After the third day, Russell played me some tapes of unissued music, some recorded by others, some by himself. He still had, in a corner of the room, the "Federal" recording machine that he had taken with him to New Orleans in 1944 to record Bunk Johnson, George Lewis, Jim Robinson, "Baby" Dodds, "Wooden Joe" Nicholas, Lawrence Marrero, "Slow Drag" Pavageau and others of the New Orleans Revival that he had largely (and literally) engineered.

During World War II he worked in a transformer factory in Pittsburgh, he said. On the night before he was due to leave for New Orleans he found that the machine, for which he had paid $225, wasn't working. But his older brother Homer, an electrical engineer also working in Pittsburgh, stayed up all night and had it running in time. In July 1944, a couple of thousand dollars (to pay the musicians) sewed into his underpants, Bill boarded the L & N Pan American for New Orleans. He took with him heavy boxes packed with the 12-inch glass (acetate-coated) discs on which the records were "cut." In all of his travels, not one of these fragile discs was ever broken, he told me.

In Canton, he actually played me a Stuyvesant Casino (New York) tape of Bunk Johnson, a Lee Collins recording he had made in Chicago, and finally (what I really wanted to hear) some unissued "takes" of Bunk Johnson's New Orleans Band at San Jacinto Hall in 1944: "Royal Garden Blues," "Yes, Yes in Your Eyes," and a long, slow blues of wonderful quality. All these I heard on headphones, as the sound disturbed his parents, who disapproved of jazz. He told me later that his mother had once complained that he had been "educated coast to coast," but had "never learned a thing." Russell also played me a modern classical record, including a percussion work that he had composed himself. But he didn't think much of it, apparently.

Before I left, Russell sold me, reluctantly, a few remaining copies of his recordings. But he refused to take more than $8 for the lot, saying that he would only have to mail the money back if I tried to give him more. In England I had paid a good deal more for bootleg copies of these discs and had spent many hours listening to them. But he wanted to give them away. It seemed he thought as little of "business" as his parents did of jazz.

There's a nice vignette of Bill Russell in the book *Drifting* by Stephen Jones, published by Macmillan in 1971. Once in New Orleans (in 1960), Jones wrote, "I even heard jazz, but that was L with the record player. She had gone into town trying to buy some King Oliver but nobody knew of him until she found a place behind shutters down in the French Quarter where she heard someone playing Bach on the violin and stopped to see the card in the window. An old man let her in [actually Bill was 55 at the time], chasing a canary into a back room with a bow. Later we got to know him and he wouldn't sell us any more records because he said that none of them were yet done as well as they should be. We pointed out what he of course knew—that the musicians were nearly all dead. In the end we had to get people we didn't know to go in there and buy them for us."

By 1965, Bill Russell was back in New Orleans. Both his parents had died. Allan Jaffe of Preservation Hall rented a U-Haul truck and drove Bill and his precious collection back to New Orleans. The family house in Canton, Missouri, was sold for $10,000. Russell moved into a fairly large apartment next to Preservation Hall, in a building owned by Larry Borenstein. He paid no rent, but was supposed to give music lessons to Borenstein's daughter. Now Russell's cardboard boxes were spread over four rooms, including the kitchen. As in Missouri, Russell shared his bedroom with unpacked boxes, jazz memorabilia, precarious pillars of accumulated research, carton mountains, disassembled violins, strings, and guitar pegs, epoxy glue, dusty newspapers unaccountably preserved, Scott Joplin sheet music, Okeh record sleeves, balls of string, boxes of tissue paper, neatly boxed Louis Armstrong and Jelly Roll Morton records (mint condition on original labels), and intriguing boxes marked "Choice." In 1966 I spent a good deal of time talking to Bill, usually during his favorite part of the day—between about one and four A.M.

Bill Russell told me that his father was a carpenter—he built half the houses in Canton—and that he was of German descent. Russell's last name is actually Wagner (and his mother's maiden name was Geigerich), but he changed it to Russell in 1929 when he began composing percussion music. In music, he felt, the name Wagner had already been taken.

When he was about 13 Russell heard jazz bands on riverboats from New Orleans, but they didn't make much impression beyond novelty. The first jazz he distinctly remembers noticing was in New York in 1928. He would hear the sound of music coming out of record stores on Broadway. "I would wonder what on earth was going on," he said. "I knew I'd heard music like that on the boats but this seemed more unusual—maybe hotter. I always wondered what the bands were—sometimes I would be on the verge of going in and asking what the record was." Later he recognized it as being music of the King Oliver, or Jelly Roll Morton type.

In 1929 he was teaching at Staten Island Academy. The kids used to play records in the lunch hour and at the end of the term they left some behind. Bill picked one up: "Shoe Shiner's Drag," by Jelly Roll Morton and His Red Hot Peppers. He thought it a "funny sounding title," so he took it home and played it. "It sounded so wonderful that I didn't get over it," Bill said.

A couple of years later he began collecting "hot" records, becoming a legend in the field. All the major collectors knew him and dreaded the news that he had been in the neighborhood. In 1934 Bill joined the Red Gate Shadow Players, a group that played Chinese music and put on oriental dramas. Russell, dressed in Chinese robes, demonstrated the various musical instruments, having studied them in New York's Chinatown—one of the least documented and more remarkable episodes in his career. On a good night, at a college, the Players might earn $100 between them—good money in those Depression years. The group played in all 48 states except Idaho and Montana, and appeared at the San Francisco World Fair in the summer of 1939. They also played at the White House for Franklin D. Roosevelt's children.

Most of Russell's "hot" collecting was done during the group's vacation in the summer. He would drive all over the country by bus, sleeping in the bus at night, and start sorting through records at Salvation Army stores, furniture stores and the like at 8 a.m., working non-stop till dusk. He could go through 20,000 records a day, eating a few bananas on the job (they could be eaten with one hand), picking out the occasional valuable record in the pile. This required vast, arcane knowledge of label colors, even sometimes label sizes, stamper numbers, pressing codes, serial numbers, and titles recorded.

The main problem was getting the records back to New York. They weighed half a pound each, which limited him to 200 records in specially constructed cases. He bought most records for 10 cents each, later selling them at Steve Smith's Hot Record Exchange for 35 cents or thereabouts. He and Steve Smith sold too soon. By the end of World War II, the best Morton and Armstrong 78s were selling for $25.

Russell had a legendary collecting success when George Beall from Detroit—a serious collector of early Paramounts, Okehs, and Gennetts who carried thousands of record serial numbers in his head—told him of a store where there was real treasure. He had found it just as the store was closing and he had to leave town the next day, so the rare items were no doubt still there. But Beall knew if he told Russell the address he would immediately head for the Greyhound bus station, no matter how long the journey, and buy up everything. Russell asked him for a clue.

"Well," said Beall, cautiously, "it's west of the Mississippi—in a basement."

Sure enough Bill went straight to the bus station and actually did find the place, in Omaha—in a furniture store called Johnson's. On his way back to New York, Bill stopped off in Detroit and told Beall of his good fortune in Omaha.

He and Steve Smith started the Hot Record Exchange in 1936 and closed it in 1940, Smith paying off Russell his share of the profits, which came to $25. Russell and another big collector named Hoyte Kline had an agreement that whoever died first would give his collection to the other. Kline was killed right at the end of World War II. Russell then auctioned off his 6000-record collection. The records were sold when "hot" collectors were still paying very high prices at the end of the war. After the legal and "bootleg" issues of the late 1940s, and then the extensive LP reissues of the 1950s, prices dropped dramatically. Bill realized "many thousands of dollars" from the Hoyte Kline collection, using the money to start his own American Music label in 1945.

Bill started composing music himself in the late 1920s. He won a Chicago *Daily News* contest with a piece called Symphonic Caprice, patterned after a movement of a Mozart violin sonata. Mostly Bill wrote percussion music—he thought there was less competition in this field, and therefore his work had a better chance of being performed. He had a few pieces published, but he finally decided it just didn't make sense. The music he was writing was getting so complicated that no one could play it. The real awakening for him came at the Three Deuces in Chicago, when Baby Dodds played solo drums at the intermission. "He'd sit there and play things I wished I could have written," Russell said, "and he was just doing it for the fun of it. It was more interesting that anything I could write—to me anyway."

Later Russell wrote some pieces that he and John Cage performed at Mills College, California, at the time of the World Fair. Bill could hardly play his own music, involving a piece of marked string attached to a pendulum that had to be lengthened with every beat . . . or something like that. Even at that stage, Bill thought, Cage "was beginning to get more interested in the lighting than in the music." Bill's interest in music as a topic of live inquiry and research switched more and more from classical to jazz.

American popular musicians almost always seem to be artistically at their peak when they first come to the public's attention. Thereafter everything tends to go downhill. Bill Russell was likewise of the opinion that the great period of New Orleans jazz was at its beginning, in the first few years of the 20th century. I asked him more than once: If he could return to the city by time machine to make recordings, what period would he choose? He said the years 1900 to 1905. That had been the Golden Age. Bill often gave the impression that, musically, everything had been in a state of decline since then. We don't really know, of course, because the first jazz record wasn't made until 1917 and the first "Negro" band wasn't recorded until 1920.

Buddy Bolden, possibly the first jazzman, flourished in the years 1900-05. Bill's researches into Bolden's meager historical remains is a story that has never been told. In an unexpected way, acting as little more than a courier, I was able to add a little fillip to this investigation shortly before I left New Orleans in 1975.

Russell began researching the origins of jazz in a serious way in 1938. He had been contacted by a recent Princeton graduate, Frederic Ramsey, Jr., who wanted Russell to collaborate with him on a jazz book under contract with his employers, Harcourt Brace.

Russell started interviewing in Chicago in the fall of 1938, talking to many New Orleans jazzmen. A principal source was Louis Armstrong, who said anyone interested in the early history of jazz should go and see Willie "Bunk" Johnson in New Iberia, Louisiana. Louis, by then famous, had recently been on a southern tour with his band and had met Bunk unexpectedly in that Cajun town. An early trumpeter born in 1879, Bunk had been Louis's "idol." Now, apparently, he was working as a handyman at Weeks Hall's plantation, "The Shadows." (Henry Miller describes staying with Hall in New Iberia in *The Air Conditioned Nightmare*.)

Bunk Johnson responded to Russell's mailed inquiry with a series of idiosyncratic letters: "King Bolden and myself were the first men that began playing jazz in the city of dear old New Orleans, and his band had the whole of New Orleans Real Crazy and Running Wild behind it." In one letter Bunk boasted that "what it takes to stomp 'em, I really knows it yet," and it was this claim that he could still *play*, not just reminisce, that led to the "New Orleans revival," turning Bill into record producer rather than record collector.

Esquire contributor Charles Edward Smith—also helping to put together Ramsey's promised book—visited New Orleans in early 1939. He interviewed Bolden's trombone player, Willie Cornish, who provided the only known photograph of Bolden's band, probably taken around 1900. Ramsey's book, published in the fall of 1939, under the title *Jazzmen*, included a heroic, semi-mythical portrait of Bolden as the first jazzman; also by repute a barber, the publisher of a scandal-sheet called *The Cricket* (no copies of which have ever been found), a drinker and a womanizer, whose mind finally "snapped" as he "went on a rampage" during a Labor Day parade in 1906.

It is well documented that Bolden was in the East Louisiana Insane Asylum by 1907, and that he died there in 1931. Beyond that little was really known about him. Bolden remains at the center of the great mystery of the origins of jazz: that it seems to have appeared almost overnight, with no "evolution," and at the same time ragtime and the blues appeared with equal swiftness.

One night, in 1975, I went to see Bill Russell, to find out more about the Bolden story. I knew that he and Fred Ramsey had both continued to do Bolden research but had hardly seen one another or communicated since *Jazzmen* was published in 1939. In 1975 Fred Ramsey himself was in New Orleans, once again at work on the Bolden story, this time with a grant from the National Endowment for the Humanities. I became intrigued, and began going back and forth between Russell and Ramsey.

After much badgering, Bill Russell agreed to dig up his Bolden material and share some of it with me. It was about 2 a.m. Wearing his regular "Salvation Army" get-up—tatty old pullover and baggy pants, he sat in his rocking chair, peering at a "Tuorte" cello bow. He said he had paid $200 for it, but thought it was probably worth several thousand dollars. He said it was getting close to his supper time. He went over to the kitchen, where he was heating up a chicken pot pie on a stove perilously surrounded by cardboard boxes and violin parts.

Somewhere in the apartment were some notebooks with the Bolden information, written up in the 1930s and 1940s, but he didn't seem to know quite where. "Might be here," he said, giving the matter momentary attention. He rummaged half-heartedly through a couple of boxes but then gave up. "Probably never will know what's in that one," he said, kicking an unmarked box at his feet.

All this life-long accumulation had its discouraging side to Bill. He told me more than once that he just might dump the whole lot in the Mississippi River (four blocks away). He had also developed an allergy of late, making it difficult for him to breathe. After much coaxing he put aside his anti-doctor principles and was diagnosed as being allergic to dust. Of which there was a lot in his apartment.

Next he sat down behind an up-ended packing case and finally said something about Bolden. "Unfortunately there isn't much dope on him," he said, nudging his spectacles up his nose and squinting across at me briefly. "Fred Ramsey is trying to write a book about him. Don Marquis is trying to write a book about him. Gee! I wouldn't have thought there was enough for a chapter. He's been dead an awful long time now." (Don Marquis did write his book, published by Louisiana State University Press in 1978. He did heroic research, but almost none of what follows was in his book.)

By about 3 a.m. Bill seemed to recover his memory. He directed me from behind his packing case: "If I were you I'd start looking in that box over there—the Charles Smith box." I opened it and it contained typed letters from Smith to Russell, carefully wrapped in layers of cellophane. On February 20, 1939, Smith wrote from New Orleans:

"I went to see Willie Cornish, and he talked for two hours or more. He's the only living member of the Buddy Bolden band, has a good memory, has documents of his war service which prove his age, has old photo of Bolden band." He added that he was saving one or two surprises—"one that's too good to announce without brass band preceding me."

The surprise was that Cornish had recorded a cylinder with Bolden. "Blank" cylinders could be bought for the old acoustic cylinder recording machines. Like tape recorders they could record as well as play back. At the turn of the century, home recording was possible.

The information about the cylinder never appeared in *Jazzmen*. In fact, Russell and Smith didn't tell Ramsey about it for three years. What the tension or feud was between Ramsey on the one hand, and Russell and Smith one the other, never became clear to me. In 1957, however, Smith published a partial account of the cylinder story in *The Saturday Review.*

The recording had been made, Smith wrote, "to please a white friend of Cornish's, whose family ran a grocery and butcher shop in New Orleans for many years."

Smith had hopes that they would find the cylinder and issue it as a record. He wrote to Russell on April 16, 1939: "I told Cornish I'd pay at least ten bucks for the record, also asked him to keep it a secret. . . . Well, that's still a pipe dream, but what a fancy one."

Russell now felt that they had made a great mistake in not announcing the cylinder-record possibility as soon as they heard of it. Many people would immediately have enlisted in the search. As it was, when the news finally came out in the 1950s, many of their friends were inclined to doubt the whole story. Classified ads were taken out in New Orleans papers, requesting cylinders. To no avail. Today, Bill said, even if the cylinder showed up there was a serious question whether anyone would be able to identify it: there is no one left alive who can honestly claim to remember what the Bolden band sounded like. (The cylinder would most likely be unlabelled.)

One piece of information had never come out—the name of Cornish's friend who was responsible for getting the recording made. According to Smith's letter to Russell, it was "Dutch" Zahn. I made a note of it that night in Russell's apartment.

After *Jazzmen* went to press in the fall of 1939, Russell enrolled at the University of California in Los Angeles, where he took two courses in Harmonic Materials and Musical Analysis from Arnold Schoenberg, the inventor of the twelve-tone scale. Between classes, Russell interviewed former New Orleans jazzmen then living in the Watts area.

He talked to Kid Ory, the trombone player, who told Bill that "Bolden got his music by going to the Baptist church with two girls called the Bass sisters—Nora and Dora. Later Bolden married Nora and had a daughter by her. Ory remembered that Bolden would come out of the church 'swinging,' as he put it." Russell then interviewed the guitarist Bud Scott in Los Angeles (and here Bill read from a black-covered loose-leaf notebook that he had found): "He told me what an expressive cornetist Bolden was—using his hand or a cup for a mute, or playing with the valves half depressed. He could actually make his cornet talk, Scott told me. If he'd see a certain girl in the dance hall he wanted to talk to he would use his cornet to call her and you would swear it really sounded like 'Mary come here.'"

Born around 1877, Bolden would have been in his twenties when all this happened. He was not yet 30 by the time he was committed.

Like Kid Ory, Bud Scott also told Russell that Bolden went with the Bass sisters to a Baptist church where they "rocked" as they sang. Russell said this church has never been identified. But it would have been in "uptown" New Orleans, close to First Street, where Bolden lived (as is known from city directories).

In May 1940, Russell found Dora Bass Pinson, living with relatives near an abandoned mill on 110th Street in Watts. Dora told Bill that her sister, who had married Bolden, was living in Iowa. They had a daughter named Bernadine, whom no jazz researcher or historian has ever found. In 1927, she wrote to the insane asylum in Jackson, Louisiana: "I am writing asking for some information concerning my father Charlie Bolden. Please inform me exactly of his present condition. As I am very anxious to know just how he is getting along. Yours truly, Bernadine Bolden." The letter was postmarked Evanston, Illinois.

Dora had no pictures of her brother-in-law, nor any other documents. She would not corroborate that he took her and Nora to church. She said that he would "break his heart" when he played, meaning that he was very expressive.

Two years later Russell tracked down Nora Bass in Iowa.

"I went to the door and had my copy of *Jazzmen* handy to prove I wasn't a blackmailer or detective," Bill said, reading from his diary. "A woman came to the door and I asked her if she was Mrs. Bolden. She was stunned that someone would ask her that. Half-heartedly, she tried to deny that she had been married to Bolden. 'Who sent you here?' she asked me. Her sister, I told her, but she kept asking me, 'Who sent you here?' She was shocked that someone had finally caught up with her. I showed her *Jazzmen* and she started looking at the pictures carefully to make sure that she wasn't in any of them. She looked especially hard at the pictures of Louis Armstrong and his mother and sister. Of course, you sometimes hear the rumor that Bolden was the actual father of Louis Armstrong, but I don't think there is anything to it. Nora was remarried and she explained to me that her husband knew nothing of Bolden and her life with 'those people,' implying that they were disreputable. I stood on the porch all the time I spoke to her. She kept looking over my shoulder because she was expecting her husband to come home for lunch.

"She had no pictures, records or anything. Of course, I was still hoping that we would find the Bolden cylinder then. But she was positive that Bolden had never recorded. 'Nobody made records in those days.' Which is not true, of course. There were Edison soft-wax home recording machines in Bolden's time. She may never have known about it, of course. It may have been before she knew him.

"I didn't dare ask her too many questions as she was ready to slam the door on me. She said that Bolden's family were 'funny people,' and she couldn't get along with them. My conclusion is that she cleared out of there soon after Bolden went crazy. Her principal memory from that time was that he was 'afraid of his horn.' He didn't run a scandal sheet, and was not a barber, but he drank a lot, she said, and he spent time at barbers' shops, which were more or less like men's clubs in the black community. 'He never made a nickel except by music,' Nora said. His mother had never made him work as a kid and always provided the best for him. She implied that he was a fine musician, well able to earn his living that way." Russell thought all this tended to corroborate other evidence that Bolden could read music and was far from being the musical primitive that some have imagined.

When Russell left she asked him not to come back, and he promised not to bother her again. Nor did he ever tell anyone else where she lived. If Nora Bass Bolden is alive today she could not be less than 100 years old. No one other than Russell ever interviewed her.

A week later I went to see Fred Ramsey, who was staying in a rented apartment uptown, close to the Mississippi River. First we went out to get some shrimp and beer from a seafood place on Carrollton Avenue. Fred told me this was close to Lincoln Park where Bolden had played 70 years earlier. Today the park is built over. A tall man then in his early sixties, Fred was wearing baggy pants and tennis shoes. He told me that he was on a very tight budget; evidently life had not been easy for him since the publication of *Jazzmen*. But he was very enthusiastic about his Bolden project.

He told me that in 1954 he had visited New Orleans while on a Guggenheim Fellowship researching Delta blues. He had gone to Bolden's neighborhood uptown and had simply approached elderly people on the street and asked them if they remembered Bolden. (It's curious to note that Russell, although he lived in New Orleans from 1956 onward, never did any such original Bolden research in the city, while Ramsey's only original research was in the city. Both, it seems, became far more active away from their home turf.) Ramsey looked down at some 1954 notes in his lap and said: "The first person I stopped said I should go to a place called Uncle Tom's Cabin. It's sort of a club, or bar, and it's still there today," Ramsey said.

From there he was directed to a man called Edmund Wise who was 66 years old and remembered Bolden well. Paraphrasing from his notes Ramsey then repeated the following story, told him by Wise: "Back in 1924 Wise and another man named George O'Leary, a singer, got to talking about old Buddy Bolden, then still at the asylum in Jackson. Having nothing better to do, Wise and O'Leary decided on the spot to go and visit the legendary trumpeter, who would then have been only 47 years old. Evidently they brought with them a gift of cigarettes for Bolden. The first thing the superintendent did was ask them not to bring up the subject of music. Well, they met Bolden and he asked about some old acquaintances back in New Orleans. In particular he asked about a white man called Oscar Zaurn . . ."

"What was that name?" I asked.

"Well, that's phonetic," Ramsey said. "I didn't know how it was spelled. But I had it spelled Z A U R N. But now we have found out from the city directories that there was a Zahn living here and that he lived . . ."

"Oh, Fred!" I cried out.

". . . he lived around the corner from Bolden . . ."

"Fred! Do you know who he was?"

Fred was as excited as I was. He stood up, tugging at the waist of his pants.

"No. Who?"

It dawned on me: Ramsey had had this name Zaurn in his notes for over 20 years, Russell had the name Zahn in his Charles Smith file for over 30 years—and the two of them had never compared notes.

"Fred," I said, standing up myself. "Zahn was the man who recorded the Bolden cylinder!"

"My God!" said Ramsey, literally clapping a hand to his forehead.

I told him about the letters from Charles Smith, sent to Russell (who was actually sharing Ramsey's apartment in New York when he received them). Ramsey had never seen the letters.

The "fix" on the name Zahn was exhilarating—two pieces of a hopeless puzzle neatly and unexpectedly fitting together. The one man that Bolden had asked specifically about, so that Wise remembered the request for 30 years, was evidently the same man that Bolden's trombonist, Willie Cornish, said they had made the cylinder recording *for*. Perhaps Bolden was wondering, in his long, wasted years in the asylum, if the cylinder had survived? No, he would hardly have thought of that. It is also inconceivable that he could have guessed that he would become a legendary figure within a few years of his death. The founder of jazz! He could never have imagined that.

But the cylinder is lost. So, perhaps, Bolden's reputation is secure.

A New Orleans newsman, Tom Sancton (father of the young clarinetist I met on my first day in the city), added to the Bolden saga. Anyone who has read A. J. Liebling's book *The Earl Of Louisiana* may remember Sancton as the reporter who drove Liebling around the state, filling him in on the Byzantine details of Louisiana politics. In 1951, when working as a reporter for the New Orleans *Item*, Sancton visited the Jackson asylum and published an interesting account, "Trouble in Mind," in *The Second Line*, a little-known jazz magazine. Sancton managed to get his hands on Bolden's medical record (and Ramsey later obtained a photo-copy of it). Sancton quotes from the pitiful correspondence between Bolden's mother and the superintendent.

March 28, 1930: "Dear Madam: Replying to your letter relative to your son, Charles Bolden, we are pleased to inform you that he continues to get along nicely. While on the ward he insists on going about touching each post, and is not satisfied until he has accomplished this task at least once. He causes no trouble and cooperates well."

After a "routine psychiatric examination" in 1925, Dr. S.B. Hays described Bolden as follows: "Accessible and answers fairly well. Paranoid delusions, also grandiosed. Auditory hallucinations. Also visual. Talks to self. Much reaction. Picks things off wall. Tears his clothes. Insight and judgment lacking. Health good. Negative blood. Looks deteriorated, but memory is good. Has a string of talk that is incoherent. Hears the voices of people that bothered him before he came here. Diagnosis: Dementia Precox, paranoid type."

Bolden was buried in a pauper's grave in New Orleans: no tombstone or grave marker, no known location, no entry in the ledger.

Russell had many depressing stories about the frailty and impermanence of the documents he had come across—or not been able to lay his hands on. Almost everything gets thrown away. It's surprising that anything survives. For example the music of Scott Joplin was published in the 20th century, Russell pointed out, and many of his rags were big hits. None-theless, some of this published and widely disseminated music almost disappeared before the ragtime "revival" in the early 1970s. When the classical pianist Mrs. Vera Brodsky Lawrence put together Joplin's col-lected works for the New York Public Library edition, she found that

Russell had the only known copy of one work ("Pineapple Rag"). Joplin's opera "Treemonisha," was published, but when I spoke to Bill he thought there were only eight copies in existence, of which he had two. There are recordings by King Oliver, Jelly Roll Morton and other outstanding musicians that survived in only a handful of copies, perhaps only one in a few cases. And this attrition is after only a few decades, in a country where there has been (by world standards) very little civil disturbance and no war.

As for the music itself—the music that had the whole of New Orleans Real Crazy and Running Wild behind it—it was played extemporaneously and never written down, so nothing beyond the recordings has survived: nothing by Buddy Petit (thought to have been the equal of Louis Armstrong), nothing by Chris Kelly (not even a photograph), nothing by Manuel Perez, and many others. It's all lost forever. The first, crucial, 20 years of jazz are a blank.

Bill clearly was worried about the fate of his own collection, containing a large number of unique items. One day in 1973 a visiting Japanese musician told Bill of a rumor that the Tulane University Jazz Archive (set up with Ford Foundation money, and with Bill as the first curator) might be moved to the music department. Bill was stunned by the news. He regards libraries and archives as insecure anyway—things get stolen, lost, moved to damp basements, broken, and above all discarded by curators with a perennial desire to winnow.

For all these reasons, Bill tried not to think about the fate of his collection, and tried to avoid his apartment, which brought on itches and inflammations. He sat for portraits by his friend, the artist Noel Rockmore, he repaired violins or any other instrument the musicians would bring him. To them he behaved with what can only be described as saintliness. At the same time, I believe, he tried not to face his assembled lifetime's labor. Editors would come and there would be talk of a book— one about Bunk or Jelly Roll Morton or Manuel Manetta or Baby Dodds or Mahalia Jackson, for he had spent years with her, too. But in the end, and after many visits, the editors would give up trying to cajole Bill and there would be no book.

It could be said that Russell himself has been defeated by his own lifetime's work. Collectors often discover that their laboriously accumulated labors become oppressive. The physical material weighs down on the spirit of the collector whose heart, in the happy hunting years, had leapt for joy when that elusive "want" was found in a dusty corner. But then the day comes when all the gaps are filled and the completed collection itself gathers dust—stirring no emotion other than apprehension or perhaps disgust in its owner.

Most of the time, Bill is one of the most unmaterialistic people you will ever meet. He owns no conventional furniture or property (I think he inherited some small amount of AT&T stock from his parents); he doesn't smoke; he doesn't drink alcohol, coffee, tea, Coca Cola or anything with caffeine; he rarely goes to restaurants and almost never travels in taxis; in the late '60s he told me that he could live on a dollar a day, and did. (In the Depression he sometimes used to walk as much as 30 miles a day to and from places of temporary employment in New York, and he said at that time he lived on 30 cents a day.) When I knew him, he would take a complex sequence of buses out to the suburb with the least expensive grocery store—more to reward the grocer than to spare his own purse. His own time he valued at nil, as he always let you know when assessing the profit and loss on his record business. (In the end he made a profit, despite his attempts not to, because a Japanese businessman showed up one day and paid him handsomely for the Japanese rights. Earlier, European rights had been bought by Storyville. American record companies have shown little interest.)

Sometimes I used to see Bill sitting in the carriageway of Preservation Hall, drowsing over copies of *The Daily World* and I think one or two other socialist or outright Communist newspapers. He would start to read an article, but he would almost always nod off, I noticed. He told me that a friend sent them, and as one might expect he kept them in a dusty pile in his apartment, along with a pile of unaccountably preserved *Times-Picayunes*. ("Might read them some day," he said, without too much conviction.) Bill had no more hopes for socialism than he had for capitalism. He was equally severe on all politicians without exception— not so much for their programs as for their ambitions. He had a very low opinion of Roosevelt, for example. The man had money, Russell said. What did he need to go into politics for? He could easily have spent his days "clipping coupons."

Bill was to me somehow a very American figure—unpretentious and knowledgeable without being worldly-wise. He had high moral standards and yet he said he had no religious faith. Still, he liked to repeat the testimony of Mahalia Jackson and others, that religion was the source of all their musical inspiration and energy. Politically, too, he was a paradox. As with so many Americans he made the most of his own liberty and independence, and treasured it, yet he seemed not to appreciate that the right to be left alone is respected only in a capitalist country.

As for the mysterious origins of New Orleans jazz, all is not lost. One night I was in his apartment looking over his important and valuable Jelly Roll Morton collection—letters, documents, precious manuscripts and many other treaures. Editors, archivists and musicologists have pleaded with Bill to deliver a book—so far to no avail. As I walked down a narrow corridor between two shelves filled with boxes and papers I accidentally kicked a loose brick. It slid noisily along the floor boards.

It was a hot summer night and Bill was wearing an undershirt, peering intently at a violin through a magnifying glass.

Without looking up he said, "Mind that brick."

"What's it doing here, Bill?" I asked. How would he ever get the Jelly Roll project off the ground if he was worrying about stray bricks in his apartment.

"Gee," he said. "My bird would have known better than to ask a dumb question like that." Then he asked me if I had heard of Albert Einstein.

"Einstein?"

He said that there had been many "dumb Englishmen" come through New Orleans, but that I was perhaps the dumbest.

Okay, I said, I give up. What's the connection between Einstein and the brick. He said he had been walking past one of the old dance halls where Bolden used to play—I think it was Longshoreman's Hall. A demolition crew was tearing it down.

"So you brought a few bricks home with you," I said, implying that his collecting mania was beyond all reason.

The dumb Englishman didn't understand. "Einstein says you can turn matter into energy," Bill said, looking up at me at last. "Maybe the reverse is true. Maybe Bolden's notes were in some ways preserved in the bricks. One day the scientists may figure out how to turn it back into sound."

Now I carefully picked up the brick. The Bolden Cylinder no doubt was lost. But here was the Bolden Brick. "Don't you think we should label it and put it in a box somewhere?" I suggested. "Otherwise, when the time comes, no one will know what it is."

I looked over at him but he had his magnifying glass up to his eye again. Something about the violin had caught his attention.

Carrying Coals to Bourbon Street

"In the matter of Jass," the New Orleans *Picayune* opined in a 1918 editorial, "New Orleans is particularly interested, since it has been widely suggested that this particular form of musical vice had its birth in this city—that it came, in fact from doubtful surroundings in our slums. We do not recognize the honor of parenthood, but with such a story in circulation it behooves us to be the last to accept the atrocity in polite society, and where it has crept in we should make it a point of civic honor to suppress it. Its musical value is nil, and its possibilities of harm are great."

In 1971, two days after Louis Armstrong's funeral in New York, a ceremony honoring the great trumpeter was held on the steps of New Orleans's City Hall. Mayor Moon Landrieu, addressing a crowd of 10,000, remarked that it had been his civic honor to represent the people of New Orleans at the funeral. Rather touchingly, Mayor Landrieu commented that he had not realized until then just how great Louis's influence had been, nor how revered was his name throughout the world.

Recently New Orleans put on a Jazz and Heritage Festival, produced by jazz-festival entrepreneur George Wein. There are plans for a park in the Treme section of the city to be named after Louis Armstrong, who was born in those "doubtful surroundings" where today the domed stadium stands. Now we have a burgeoning "Cultural Center." We have a Jazz Archive at Tulane University and a Jazz Museum in the Royal Sonesta Hotel. Culturally, jazz has definitely arrived in New Orleans. Jazz is respectable at last.

"Jazz has become infinitely more respectable since I first came to New Orleans," said Richard B. Allen, the portly curator of the Tulane archive,

34

who came to the city in the late 1940s. "For one thing, it has become academically respectable. Secondly, people are beginning to realize its attraction to tourists."

Justin Winston, the curator of the jazz museum, agreed with Dick Allen. "There was a survey not long ago," he said. "They asked tourists to put down what they came to New Orleans for, and jazz was third. I think the French Quarter was first, and the second may have been food. Jazz, in fact, may be the most important attraction, because these categories overlap. For instance, how many people go to the French Quarter to hear jazz?"

The irony is that the sudden arrival of jazz as a tourist attraction, as a festival package, and as a civic virtue, may well be coinciding with its demise. Jazz, according to one unpopular view, has already been here and gone. As with Acadian French, a vestige remains and so there is a belated interest in preserving it. Certainly there is something odd about the idea of a jazz festival in New Orleans. Are we now reduced to carrying coals to Newcastle?

Like so many other activities belatedly recognized as "cultural," jazz is now regarded as a suitable recipient of grants, donations, federal funding and corporate underwriting. This only adds to the uneasiness of the pessimist school. After all, King Oliver and his Creole Jazz Band didn't need a grant to take jass "up the river" to Chicago. In today's climate they would not have proceeded an inch upstream before their expenses were covered by the Schlitz Brewing Company, or were certified by the National Endowment for the Arts as an "outreach program."

The problem with earlier jazz festivals in New Orleans is that the black population stayed away in droves. This year George Wein shrewdly included rock stars in the auditorium concerts—big names with little or no connection to the jazz history of New Orleans. The idea was that the blacks would come to the concerts, and once there, the pied piper of advertising and "word of mouth" would lure them to the "cultural" events. It seems to have worked fairly well, at least as far as the box office was concerned.

Allison Miner, an assistant to George Wein who also helped put on last year's festival, was dismayed that the Staple Singers came to town last year "a week after the festival was over." And yet they sold out the auditorium. "I saw Pops Stapleton on the street and I told him I'd like them at the festival." The Staple Singers (who now call themselves "soul" singers rather than gospel singers) were duly booked. Allison Miner explained that successful auditorium events can underwrite the cost of the Heritage Fair—a four-day event with five music stages, a gospel tent, food stalls and Louisiana craft exhibits which many people regard as the most attractive aspect of the festival.

It has not escaped the notice of some members of the New Orleans jazz "establishment" that this procedure involves a mild distortion of the word jazz. Allan Jaffe, the manager of Preservation Hall, called most of this year's auditorium concerts "blatantly commercial. If they think that putting on Stevie Wonder or Howlin Wolf is helping New Orleans jazz, they are naive."

"They *are* naive," added Larry Borenstein, who rents Preservation Hall to Jaffe. The "they" whom Borenstein indicts includes the Jazz & Heritage Festival producers and their board of directors. "They don't *like* jazz," Borenstein grumbled. "They like Viennese waltzes. They might as well get Herb Alpert and the Tijuana Brass or Henry Mancini to come down here." (Actually, Herb Alpert "sat in" from time to time at Preservation Hall and has done so very effectively, not merely adapting himself to the ensemble style but seeming to have a better understanding of it than the old-timers he was playing with.)

"The question is whether New Orleans needs a jazz and heritage festival," said Jaffe, who admitted that it was not entirely by accident that in recent years he and the Preservation Hall Jazz Band have been out of town at festival time. (This year the band, along with the Benny Goodman Sextet, played at the auditorium on Wednesday.) "The answer is," Jaffe responded to his own question, "we do need the fair and we don't need the auditorium concerts—not if the people they are going to feature don't help jazz in New Orleans."

"We cannot survive like that," Allison Miner responded. "we need the auditorium concerts because, among other things, they help to pay the cost of the fair. They're really going to cry next year when we have a rock 'n' roll boat ride with Jerry Lee Lewis on the Steamer President. No—I'm kidding. I don't know whether we are going to or not. But I'd like to. Jerry Lee Lewis is from Louisiana, you know. He's a part of Louisiana's heritage."

So it all boils down to economics and the imperative of financial survival. Give purist jazz concerts (as they did in earlier years) and the audience won't come. Compromise a bit, extend the definition of "heritage," give top billing to a soul singer or two, and the accusations of commercialism and historical irrelevance begin to fly.

The doleful fact remains that it is jazz itself that has become largely irrelevant. Its functional role in the life of the city has all but disappeared.

Jazz has been reduced to a piece of cultural icing on the urban cake. It was not always so. Back in the days when jazz was jass it was so embedded in the life of the city that the *Picayune*'s editorial barbs could have made no conceivable difference (not that any jazz musician then playing would have read the newspaper).

"In Buddy Bolden's day they didn't have to worry about people promoting jazz," said William Russell, the jazz historian and collector who frequently helps out at Preservation Hall in the evenings. "Bolden didn't need a grant because he had so many jobs that he would have to get other trumpeters to go out and play in his place," Russell said. "In the old days in New Orleans, jazz was used for everything from birthday parties to funerals. They played music for excursions and they played for dancing, for banquets and ball games. In Hollywood they use searchlights for store openings. But here the standard thing was to have music. Sometimes they still do even now. Bands used to play to advertise everything from prize fights to the dance that night.

"In the old days there would be picnics at Milneburg—bands would go out to Lake Ponchartrain on the Smokey Mary which ran alongside the Elysian Fields. There would be music for every kind of parade and, of course, they played on the lake steamers. They might have had music on some of the packets that went up the Mississippi.

"Today the music is used for concerts," Russell continued. "It's used for festivals, and it's used on Bourbon Street. All the wrong things, in other words. It's used for juke boxes, for recording and for television—all the wrong things, because music is supposed to be for dancing and having a good time, and not just sitting there. But anyway, I'm supposed to cooperate with the jazz festival and George Wein is a nice guy.

"Another thing that's been bad for music here and everywhere else is electricity," Russell said. "One band over the radio is going to take the place of a lot of other bands. They used to have a band playing on every corner practically. But they don't need that any more. Besides that there are some social changes. Dancing is not very popular any more. I went to Filmore West when Billie and De De Pierce were playing there in 1968. A couple of rock 'n' roll bands played first and I was surprised to see that nearly everyone there just sat and listened—even with those drums banging away and all. So there's something different there. But I don't worry about it because I think tom cats know more about sociologizing and philosophizing than human beings."

Bill Russell himself received a grant from the National Endowment for the Arts to work on a variety of jazz research projects, including a

compilation of Jelly Roll Morton documents. "Better to use the money
that way than to use it to drop bombs on the Vietnamese," he commented. Still, he conceded, Louis Armstrong and Bunk Johnson and Jelly
Roll Morton himself did not need grants.

Jazz Museum curator Justin Winston added: "They don't give grants to
people who might turn out to be good. They give them to people who
have already proved themselves in some way. If they had had grants in
Louis Armstrong's day, he wouldn't have gotten one until he started
playing in the big bands in the 1930s. By then he had already made it.
The grant system can't work any other way."

Meanwhile, this year the Tea Council of the USA has combined forces
with George Wein "to launch a nationwide talent search for youthful
musicians" who will be "selected to play and sing side-by-side with
veteran jazz greats" at the Newport Jazz Festival. And at Southern
University, the Jaguar Jazz Club '73 announces that one of its objectives is
"To enrich the cultural life of the State of Louisiana through jazz and
related musics." A priority: "To maintain throughout the school calendar a
viable jazz-oriented performance series utlilizing student, community
and faculty resources." Let's hope it stays viable.

Amidst all this cultural proliferation, the archive has been neglected.
The Jazz Museum (its very existence seems to sound a death knell) will
soon have to leave its home in the Royal Sonesta Hotel. It operates on a
tiny budget, while its affiliated archive, now being moved into a new
location on Euterpe Street has just about no budget at all. The city is
supposed to be restoring Perseverance Hall (near the new Theater of the
Performing Arts) as a final resting place for the Jazz Museum, but that
will take at least another three years. In terms of damage, "three moves
equals one fire," Winston said.

"People in this city don't think of New Orleans music as anything
unusual that needs to be saved, so why bother?" Winston continued. "In
maybe 10 or 20 years it won't be here any more, so maybe this is our last
chance to find out the way it was. Very few people in New Orleans
bothered to save anything. You can count on the fingers of one hand the
people who collect photographs of old jazz bands. Outside New Orleans a
lot of people do, of course."

Meanwhile the Louis Armstrong Park and the Cultural Center will
grow apace in the Treme. Before this is heralded as a stroke of civic
enlightenment, it is worth recalling that the Treme section that was
demolished to make room for culture was itself one of the most important
places where jazz was played. San Jacinto Hall, where Bill Russell

recorded Bunk Johnson in the 1940s, and where many of the best jazz-men played for dances in the early years, stood on the very site of the Theater of the Performing Arts. Thus indigenous culture is replaced by a more standardized version, imported from New York and from Europe. In that sense the theater stands as an inadvertent mausoleum for New Orleans jazz. An award-winning group will no doubt soon enough perform there as an exercise in cultural "enrichment."

When jazz is admitted finally to the culture centers, the cultural tradition that produced it will have ceased to exist. When it is deprived of everything but its residual aestheticism, jazz must finally come to rest in the musical museum of the concert hall. It is probably always true that art forms have most vitality when they are least suspected of being artistic. And that, in New Orleans's case, was just about the time the *Picayune's* editorial appeared. Louis Armstrong's career was then just beginning.

Larry Borenstein

Most days of the week E. Lorenz Borenstein may be found sitting behind his desk in his art gallery at 730 Royal Street, a building he bought a year ago (in association with Allan Jaffe, the manager of Preservation Hall), for $200,000. A rotund figure in his early fifties, Borenstein is nearly always to be seen in one of his large collection of T-shirts and a grotesquely crumpled pork-pie hat. "It's a non-kosher hat," Borenstein admits, but he hangs on to it if only because it has a signed ink drawing by the artist Noel Rockmore on the inside of the crown. Not that that qualifies it for the Louvre, Borenstein allows, but he'll consider an offer all the same.

On the afternoon I went to see Larry Borenstein, he was behind his desk as usual, watching the half-dozen or so customers browsing through the pre-Columbian figurines and other artifacts on display in his gallery. The walls of the gallery were panelled with heavy poplar planks rescued from old Mississippi river barges—typical Borensteinian decor, combining parsimony with good taste. It is also to be seen in Vaucresson's Cafe Creole on Bourbon Street: Recycled Rococo, perhaps, or Borenstein Baroque. Hanging from the walls are paintings by Noel Rockmore, Charles Richards and Sister Gertrude Morgan, a New Orleans primitive.

"I regard myself partially as a collector," said Borenstein, who may also be defined as an investor, man of property, salesman, entrepreneur, as well as a one-time philatelist, illusionist, rare-book dealer and currency speculator. "The buildings I own are a part of my collection," he said. "Each one I regard as an irreplaceable work of art in its own right." Larry, who was wearing a turquoise suit of recent vintage and a faded salmon T-shirt, began to describe his property holdings in the French Quarter, more than once alluding to his "portfolio" of buildings.

"Frequently I've turned down buildings that I don't want to add to my portfolio for purely aesthetic reasons," he remarked. "When that happens I frequently refer the vendor to someone else I know who may want to purchase the property if it's a good potential investment—Jules Cahn, for instance, or Allan Jaffe.

"I am an investor, and yet they attempt to attribute to me the motives of a speculator," Borenstein went on. In conversation, one soon notices, he has a way of responding to the criticism of anonymous detractors who persist in misconstruing his motives despite his patient efforts to set them straight. "In a speculation you have a completely different set of criteria: minimum equity and a short hold. There have been some fairly successful French Quarter speculators—Peter Ricca, for instance, or Bill Copping or Sam Miceli. But speculation is not my objective."

He nudged his horn-rims up his nose and hooked one ankle over a knee before continuing: "I have always paid the asking price for the properties I own," he said. "I either pay the asking price or I don't buy at all. If the building is worth having, the price doesn't make much difference. Look at it this way. In the whole world there are only 12 Royal Street blocks, therefore 48 Royal Street corners. Of these, approximately half will never come on the market, and the remainder only occasionally—say once every five years. If you have the opportunity to buy a Royal Street corner—and there may be only 12 such opportunities between now and the end of the century—what difference does it make if you pay $150,000 or $200,000?"

This philosophy has worked well for Borenstein. He bought 624 Bourbon Street (Vaucresson's Cafe Creole) in 1965 for the asking price, $110,000, while the nearby competition, who wanted the building for an expansion of their facilities, lost out by bidding only $80,000. Today Borenstein reckons the building is worth double what he paid. He is also convinced that the price of French Quarter property will continue to increase.

"I think it's still considerably underpriced," he said. "When you think that prime Bourbon Street footage sells for about $5000 a front foot and Canal Street for $10,000 a front foot, you can see why. You can project that Canal Street values will definitely drop because of the Poydras Street development, and that Bourbon Street will go higher. At $40 per square foot at present, prime Bourbon Street property is probably valued less than a shopping center. Bourbon Street property was built to last, while modern structures are built with planned obsolescence. Add to that, on Bourbon Street you have a whole new clientele of tourists coming in waves every two weeks or so. . . ." Borenstein paused and reckoned with the financial possibilities.

"I have confidence in the Quarter," Larry said finally. "And I demonstrate that confidence by investing in it. I am willing to pay a fair price or even more for a well-located building which appeals to me aesthetically. Under those circumstances I'll be glad to add it to my portfolio."

The telephone rang and Larry's hand darted at the receiver as though the caller might hang up before the second ring. He briefly exchanged a few words with his caller.

"In my case I'm a fat old man with kids," Borenstein said, lighting up a Kool with a quick pass of his lighter and giving his glasses another thumb-nudge. "I can't buy insurance worth having. But I'm in the taste business. I consider I have trained taste, and so I'm not looking for a quick turnover. What I'm interested in is a long-term investment. Some people paint their property up and try to make a quick sale. Better, I think, to repair the roof and make it functional in small units, a bit at a time."

This remark is in apparent response to some of Borenstein's critics who complain that he is a disguised slum landlord who displays more interest in collecting rent than in making repairs. The rents tend to be high, too, his critics say. Not so, Borenstein replies—they were formerly too low. "Many estates with absentee or disinterested owners," Borenstein explained, "have rentals based on their cost of acquisition. If such properties change hands, then rents must go up."

Borenstein defends Royal Street rents of $1,000 or so a month with an argument that runs like this: One thousand dollars a month is $35 a day. A prospering Royal Street business with a good inventory can take in $300 a day, and therefore rent is only ten percent of the gross. "Rent is one of the least expensive items in the cash outflow," Larry believes. Here's the clincher: What's the difference between $400 a month and $1000 a month? "Seven thousand a year, and a business doesn't belong on Royal Street if an additional $7000 a year is a financial burden which can't be borne."

A well-dressed couple walked into the gallery—likely prospects I thought, and so indicated to Borenstein. He glanced at them in quick appraisal and shook his head. He guesses right 80 percent of the time, he reckons. He was right in this instance. They walked out without buying anything.

The New York artist Noel Rockmore (featured as a child-prodigy painter in *Life* in the 1930s) came into the gallery with a girl on his arm. Borenstein owns approximately 1000 of Rockmore's paintings, drawings and sketches. He eyed Rockmore and friend in thoughtful surmise, and flicked his zippo lighter up against the tip of a fresh Kool. On this occasion he kept his thoughts to himself.

During one phase of Borenstein's immensely varied career he worked as an assistant to the illusionist Joe Dunninger. "I received his thought waves," Larry said with a straight face. I was reminded that Larry once described his art gallery operations to me as being divided into two parts: downstairs the reception, upstairs the deception.

Born in 1919, the son of a Milwaukee merchant and investor, a "chess prodigy" at the age of ten (he claims), Borenstein left Milwaukee at the age of 14 and landed a job with the "Streets of Paris" sideshow at the Chicago World Fair. Sally Rand did the fan dance that year, and Larry Borenstein, billed as Prince Cairo, worked up a fortune-telling routine. Then came the stint with Dunninger, followed by a period with the Royal American Shows carnival on the road. Borenstein operated a sideshow unofficially called Pickled Punks (officially, Mysteries of Life), featuring fetuses in various stages of development and deformation. The choice exhibit had two heads, at least one made of Latex, Larry concedes, but convincing nonetheless. Borenstein lectured on embryology to the gaping audience.

At the end of the season he turned his hand to the magazine subscription business. Larry did very nicely touring Oklahoma and Texas during the summer and fall of 1937. He introduced a notable innovation—towing a rack of bicycles behind his station wagon. Kids he had hired would ride up on bikes to suburban front doors soliciting subscriptions, creating the illusion that they were from the neighborhood. Borenstein says he cleared about $15,000 in nine months—"phenomenal for the Depression," he allowed. "I've made money at almost everything I've done."

He put himself through Marquette University with the proceeds, majoring in philosophy. At the same time he worked for the Milwaukee *Sentinel*. Then he was a reporter for the Toledo *Blade*. Borenstein was working for the American Vacation Association and was on his way to Florida when he first arrived in New Orleans. The association had employed him to help stimulate winter tourism in the South. On the night Borenstein arrived in New Orleans, in December 1941, Pearl Harbor was attacked. His boss advised that the war wouldn't last more than two or three months and to stick around and make some contacts in New Orleans. Borenstein stayed for more than 30 years. "The war knocked hell out of the travel business anyway," he said.

For a few months Borenstein worked as a salesman for Uncle Joe Rittenberg, "a famous pawnbroker on South Rampart Street," Larry explained with a touch of awe. Larry opened his first shop at 814 Royal, selling stamps, while an associate, Al Williamson, sold coins. Rent in 1942 was $20 a month. The same space fetches $600 today, "and I don't own it," Borenstein said. In 1943, he moved to 706 Royal ($30 a month then, $500 now).

Borenstein continued in the stamp business until about 1950, eventually becoming an editor of *Weekly Philatelic* (published in Kansas). His Old Quarter Stamp Shop and Jolly Roger Book Shop in Pirates Alley did a thriving business. Borenstein also was editor of the *Old French Quarter News* (precursor to the *Vieux Carre Courier*) in its final months. Outside his shop, artists were encouraged to set up their easels in the open air.

"It's on my conscience that that tradition, which later spread to Jackson Square, started with me," said Larry, who takes a dim view of sidewalk artists today. (Cynics suggest that this is because they compete with his gallery for the tourists' dollar. Larry's rejoinder, that what he objects to is the *quality* of sidewalk art, is less than convincing, even if he does claim to be the French Quarter's current Custodian of Good Taste.)

Borenstein said that the Jackson Square tradition of sidewalk artists sketching tourists started when he sued St. Louis Cathedral "because it seemed that painters weren't allowed to display on the cathedral railings without political help." The suit stopped the use of the cathedral railings entirely, and artists moved next door to Jackson Square. "I told the pastor, Father Burns," Larry recalled, "that I wasn't the first Jewish boy to try to drive the money changers from the Temple walls." Today, Borenstein estimates that Jackson Square sales and services (framing and so on) add up to about $1.5 million a year.

As his stamp business developed, Larry became more and more involved in changing money himself, usually for foreign sailors who were willing to accept dollars at less-than-favorable rates. Borenstein began travelling to such spots as Surinam, Curacao and Guatemala, where he would buy up substantial portions of the latest stamp issues in the hope that the government would not reprint them.

This worked fine in Guatemala, especially, where Larry happened to have a relative who was friendly with the ruler of the day, General Jorge Ubico. He would tell Larry the size of the "print run" of each issue. Borenstein woulj then buy up most of the denomination least likely to be used for postage. "Collectors who wanted a complete set would eventually have to buy them from me," Larry said, missing a beat and then adding with a sidelong glance, "at a premium. . . ."

Speculating in foreign currencies led to one of the occasional financial disasters in Borenstein's career. When the pound was devalued (to $2.80) in the early 1950s Borenstein was caught "short" with a large quantity of sterling bought on margin. As a result he found himself broke, even after selling off his entire stamp collection for a six-figure sum. Gradually he climbed back into solvency. By 1954 his bank balance was $500.

Art dealing, his next venture, appealed to Larry because it required little capital. Nestor Fruje, Charles Richards and Richard Hoffman were among the artists who left their work with Borenstein on consignment at Associated Artists Studios, 726 St. Peter Street, now Preservation Hall. Borenstein had obtained a long-term lease from the owner of the building, who lives out of town. Later artists represented by Borenstein included Xavier de Callatay, Sidney Kittinger and Noel Rockmore, each of whom was commissioned to paint a series of portraits of old jazz musicians. Despite, or perhaps because of the persistent refusal of the New Orleans art establishment to acknowledge his existence or activities, Borenstein believes he has represented the best artists in the city. Frequently their national reputations have exceeded their local reputations.

In 1957, Borenstein bought his first French Quarter building, 732 St. Peter Street (now occupied by Ciro's Flamenco Dancers) for $32,000, all of it borrowed. By this time, Borenstein was making frequent trips to Mexico to dig up pre-Columbian artifacts. He specializes in West Mexican figurines. Some that he has collected are 2000 years old, none less than 600 years old, he claims. "Most of my inventory I actually took out of the earth myself," Borenstein said. "That's what I did for a living. I was a grave-robber."

He looked at me quizzically through his thick horn rims. "What are you laughing at?" he inquired. "Of course, I don't do it any more," he added. "It's too dangerous. Law enforcement has rendered that part of my career obsolete. It's too risky to get them out of Mexico. In the U.S. government's zeal to stop marijuana from coming across the border they funded check points for the Mexican government, and it became no longer possible to slip the guard a few pesos when he was looking through your car." Borenstein said he has been to jail three times in Mexico as a result of his grave-robbing activities. "The last time, in 1969, it cost me a great deal of money to get out. They confiscated my Buick station wagon and its cargo, plus I had to pay a large cash bribe to get out." As with other reverses, Larry seemed philosophical about it.

As well he might be, with one of the largest holdings of West Coast Mexican artifacts in the United States—several thousand items, which he sells at prices ranging from $10 to $2000. He not only anticipated the collecting trend in this field but helped to popularize it with a series of winter showings at Macy's in New York. A number of collectors started buying at his instigation. Since 1966, the year of the first Macy's show, pre-Columbian art prices have doubled.

There's little doubt that Preservation Hall—a name he dislikes and would have preferred to change—has been Borenstein's most successful venture in terms of nationwide and even international acclaim. Yet the hall never spent a penny in advertising. Borenstein's natural parsimony combined perfectly with the public's preconceptions as to the authentic ambience of early jazz. A shabby decor paid a handsome dividend, in other words. The early jam sessions with Punch Miller and Kid Thomas Valentine, supported by donations from the few French Quarter strollers who would drop in, soon blossomed into the most successful presentation of traditional jazz in the city. At the beginning, in 1960, there was always the danger of a police raid if white musicians joined in with blacks.

Considering Preservation Hall, his huge collections of art and pre-Columbian figures, and his holdings of real estate, you might think that Borenstein could now sit back and rest, but that is hardly his style. He's hoping to acquire more buildings for his portfolio soon. He loves New Orleans and is optimistic about its future. "I think it's the best city in the U.S.," he said. "If I thought there was one better I'd be there. There are a fantastic number of things wrong with New Orleans, of course, but I can cope with them better here than I can elsewhere, and the things that are right about New Orleans don't exist anywhere else. As for the future, there's no reason to suppose that any of this escalation in value is going to stop. We have the domed stadium coming, supposedly 'Light and Sound' in Jackson Square in 1974, the pedestrian malls are a great asset, and generally tourism should increase now that Vietnam is over."

In a couple of hours you can only scratch the surface with Larry Borenstein. He has so many stories to tell, just as there are many about him. There was the Bohemian Art Colony at 912 Toulouse Street in the 1950s, for instance, which he helped to start. It had been occupied by the Chinese Merchant Association during the war—members of the On Leong Gang, or something like that. When construction workers dug up the yard to put in drainage, they found all these Chinese bodies. . . .

And the Mexican gold? I asked Borenstein about that rumor. It all started in 1947 or '48, he said, when he noticed a surveillance bug in a light fixture in his office. He told his secretary about it and they agreed to ad lib fanciful tales for the benefit of listening agents. They talked about gold brought in from Mexico and hidden away in the office. When they returned after the weekend the place had been ransacked.

"Now Larry," I asked, "why was the FBI interested in what you were doing in 1948?"

"Well now, that's another story. . . ."

Hippie Mardi Gras

The street people are on the move again, and at this time of the year most of them are moving toward New Orleans. They come for Mardi Gras, for the great big party in the streets that they have heard so much about in the street-people meccas of Tucson, Boulder, Berkeley and Austin.

New Orleans for the Mardi Gras sounds so enticing from the vantage point of those distant towns where the mendicants of America gather. New Orleans! Where the sun is shining, the wine is flowing, the beer is foaming and good things rain down like manna from heaven! Just a great big party in that easygoing southern town, where fun-loving people ride the Carnival floats and dispense doubloons to the eager crowds . . . free! For the hungry, there's red beans and rice at Buster Holmes's—the cheapest hot meal in the United States. At Jackson Square you can sleep under the palm trees and pluck your breakfast from the banana trees. There are free clinics and crash pads and the usual amenities of subsidized living, obtainable in New Orleans, as elsewhere, with a certain amount of attention to the rules and regulations of the Federal, state and local bureaucracies.

So New Orleans is the place to come at Mardi Gras. Twenty thousand street people might come. That was the figure one heard. But however many do come, they all share one characteristic, almost by definition: they bring with them to the Crescent City very little money.

Although the footloose life of free and easy travel is much romanticized, those who practice it find they are rarely hailed as heroes by the workaday citizenry, and true to form those itinerants who have made their way to New Orleans this year have not been the recipients of any Huckleberry Finn Awards. In fact, the 1973 flock is wondering just where the party is, exactly, and if there is one, why the hosts haven't been welcoming them with open arms. They have been greeted not by a shower of doubloons but by Felony Action Squads, policemen disguised as hippies, and the threat of instant arrest if they so much as ask strangers for a dime. Panhandling has all but ceased as a result. As this is the primary source of income for most street people, many have already given up on New Orleans and moved on to greener cities.

The browned-out turf of Jackson Square, between St. Louis Cathedral and the Mississippi River, is the principal recreation area of the street people. Here they come to sleep, or to deposit their unwashed bodies, or to gambol about if the mood so strikes them. Policemen watched warily as a large group frolicked while guitars twanged and the inevitable flute hooted last Friday afternoon, eleven days before Mardi Gras day itself. Sitting cross-legged on the grass are a young couple from Daytona Beach—Harlan and Debbie. They have a puppy with them.

"People call me 'Freedom,' says Harlan, 26. "I came to New Orleans to see the Mardi Gras. I've never seen it before, but I heard all kinds of rumors and stuff about it. Supposed to be one big party, man. I heard there's all kinds of beer and all kinds of good stuff floatin' around, and good times; wine, beer, dope, and anythin' else."

Harlan says he is a panhandler in Daytona—that's what he does for a living. Debbie, 19, "sits around on the boardwalk" when she's home. "And we drink Mad Dog and vodka and tequila and rum and Coke," adds Harlan. Debbie says she "dropped out of high school three times." Harlan briefly sampled college in Boisie, Idaho. They plan to stay in New Orleans through Mardi Gras—"unless it gets too radical or rowdy, you know."

New Orleans "is a little bit different than I thought it was going to be," says Harlan. Obviously he is disappointed but doesn't want to admit it. Debbie thinks it's "a lot different."

"The people," Harlan explains. "There's not as many around as I thought. Plus, the police will bust you if you panhandle." Nevertheless, he has been panhandling so far, without being caught.

"Eh, where y'all been," Harlan calls out to some returning friends. "D'ya get that job?"

Bill, wearing surprisingly clean clothes, says: "I like New Orleans. It's everything I expected. I'm a jazz freak, you know? The early blues?"

A young girl aged about 20, from a "small town in New Jersey nobody's ever heard of," says "I ain't rappin' on a tape recorder."

Terry Sherar, an Englishman, has been traveling across America in a van since December. He thinks New Orleans "has a nice ethnic identity to it." He is staying in a Canal Street rooming house.

Dan, aged 18, from Lanham, Maryland, is reclining on the grass reading a Jesus paper in desultory fashion. What had he heard about New Orleans and Mardi Gras?

"Nothing." Pause. "I was livin' right across from D.C. and this dude came down from Connecticut. He crashed at my apartment for a while and said he was going to San Diego, so I says well, maybe I'll go with you. He says fine, maybe we'll hit Mardi Gras together. And then we met some other guy who was heading out here, and he said he'd been to Mardi Gras and it was just a big party. I came down with five people. Two of us are sleeping wherever we can. The rest left yesterday."

Go to college?

"No, I never got out of high school."

He says he spends the day "either walking the streets or sitting in the park." His father, he says, "works in computers or something."

Is he as disillusioned as he seems?

"No, I'm just tired. I just woke up. I like New Orleans. I'd stay here if I could, but I have to go to San Diego. Maybe I'll come back."

The Jesus paper?

"Somebody gave it to me. Someone handing them out."

Does he believe in Jesus?

"I don't know. Maybe. I don't think his is the only way. I think it's all right what you do as long as you don't bother people."

The most popular eating place for street people in New Orleans is Buster Holmes's, at the corner of Burgundy and Orleans. Red beans and rice, French bread and margarine slab can be had for 35 cents. The unfailing good cheer with which Buster himself serves all comers has given his establishment an unrivalled underground reputation. Street people who have never been to New Orleans before know where to go when they're hungry.

Last Sunday, the tables of Buster's dimly lit front bar were filled with voraciously hungry youngsters, most of them with hair scraggly and matted from the rain. They wiped up every last morsel from their plates. Some wander in with no money at all and watch to see if any food is left. Buster doesn't have to carry out much garbage.

A group of street people sitting around one table included Mark, 19, from Ventura, California; Barry and Marsha from Minneapolis; and Mike from an unspecified location in California. Mark has long hair but he is conspicuously tidy and well groomed. He has no money at all. He has been roaming wherever the spirit moves him for about a year.

The main centers for the young mendicants are Berkeley, California (friendly officials); Tucson, Arizona (warm in the winter); and Boulder, Colorado (wild mountain terrain). Street people tend to stick together, Mark said, and if one "scores" a dollar or two he will share it with friends. Some people he knows came to New Orleans with 50 cents in their pockets. Others came with nothing.

Someone asked Mark if this way of life didn't involve a good deal of sacrifice—going without food, huddling in doorways, sleeping near central heating ventilators, using newspapers as sheets, and so on.

"Sacrifice?" he replied in a tone of disbelief. "I don't see where there's any sacrifice. I can go where I want to, do what I want. Sometimes I get a job. I worked here on a garbage truck one day. And I got a job from a temporary help agency as a waiter one night. Made about $16, and we could drink all the leftover booze."

"What are the good times for street people?"

"The good times are when you are hungry and someone gives you a dollar, or when you are tired and someone gives you a place to sleep, or when you score a bottle of wine or something to get you high." He spied a neglected crust of bread across the table. "Isn't anybody going to eat that bread?" he asked. Marsha handed it to him.

"You meet a lot of different people," Mark went on. "The STP family—there's a lot of them in town. You haven't met them? They're like bikers without bikes. Foot gangs. They like to drink a lot. Of course, there are problems. The hygiene of most of these people is pretty bad. Almost every chick you meet has got the clap, or says she has. And then the guys always outnumber the girls on the road. I expect I'll stop traveling around like this sometime. Maybe go to college. But I don't worry about the future too much."

Mike, also from California, mentioned that he had been arrested for panhandling and had been put in a cell with 16 other people. He was released when a friend paid his $25 bail. He said he had been arrested in this way: A hippie-type person approached him with an offer to get some wine. Great! said Mike. But the hippie had no money, which made two of them. But they set off together anyway, hoping to "score" some cash. Around the corner came a "straight," clinking coins in his pocket.

"Hey man, got a dime?" asked Mike.

And off to jail he went, escorted by the phoney hippie—a wolf in sheep's clothing.

Mike has had ample occasion to observe the resentment of the locals at his and others' penurious condition. "But the idea of Mardi Gras wasn't to make money in the first place," he said, displaying more knowledge of history than most street people possess.

Thomas, cleaner-cut than most, said he was a carpenter by trade. He lives in a candle-lit cabin that he and two friends built in Manitou Springs, Colorado. They live off food stamps, and work now and then, too. He came to New Orleans two years ago for Mardi Gras, so he knew what to expect. "If my old lady don't come down here soon I'm going to leave," he said. "I've been here a week and I haven't had a good time."

Sharon, 18, is from Tucson. A black cowboy hat covered her frizzy blonde hair. Her nails were painted with maroon varnish, and her fingers adorned with five emerald-colored rings. She hardly counts as a street person now because she has a job as a dancer on Bourbon Street and seems to have decided to stay in New Orleans. She announced that she has a plant named George Copperleaf in her apartment. Alone among the table-sitters she seemed to be floating along independently in her mind, oblivious to the ups and downs of life on the road.

The Ursulines Street headquarters of the Process Church of the Final Judgement (founded in England in the 1960s) is a haven for the homeless in New Orleans. You can't stay overnight, but they will refer you to St. Mark's Community Center run by the ABBA Foundation. There you can occupy floor space for a dollar a night. The Process Church has two discussion rooms in the back of its premises, and in recent days the rooms have been much-frequented by exhausted itinerants. The church has 17 full-time members in New Orleans.

Last Saturday a number of weary travelers sat in and between the orange-and-yellow sofas. A Cat Stevens record played softly. The Process symbol, a strangely off-center cross, hung from one wall. Keeping an eye on the mendicants that afternoon were Sister Raphaela, Brother Patrick and several other blue-suited initiates who bustled in and out, their manner indicating that no crisis could be too great for them. Sister Raphaela, recently "transferred from Chicago" to New Orleans, said she found the work "very exciting" and spoke of "energy interchange" with those who came to rest at their establishment.

She mentioned that one of their more unusual visitors that day was a man who lived with a possum in a cave outside Austin, Texas. Steve was his name. In fact, he was huddled up in a corner nearby. He wore leather-patched jeans, leather thongs, lengths of string around his neck, with a feather quill tied to another piece of leather. His hair was sticking out everywhere and his eyes were staring wildly.

"You pay no rent in a cave," said Steve. "And you have no landlord." He sighed deeply, pulling in air with effort. His pupils were dilated. "Maybe a hundred thousand people could live in the caves around Austin. There's bass and catfish in the river right below my cave, the Colorado River. All you have to do is find a town which has a good hip community, and panhandle. All you need is a sleeping bag and $5 for candles, matches, rolling tobacco and food. . . ."

He paused, drew out a can of Spar Var spray paint ("Bright Gold, Dries Hard & Fast") and squirted a long jet onto an old black sock balled up in his hand. He applied the sock to his wide-open mouth and drew in sharp, deep breaths.

"This country is so rich they can sell dope in cans," he said. Then he forgetfully repeated his formula for easy living: "All you need is a sleeping bag and $5 for candles. . . ." This time he added: "And you can work if you like. You don't have to panhandle. Or you can steal if you want to do that."

He explained that a yogi man lived in another cave "six blocks away," but that he didn't know of any other full-time cave-dwellers around Austin. Some came and went on a trial basis, he added, freshening up his sock with another lengthy squirt of Spar Var. "I have Homer and Shakespeare and Plato up in my cave," he said. "I read the great books."

Why had he come for Mardi Gras?

"The mountain people came south during the winter. You heard that, I guess?"

The spray paint? What does that do?

"Well, . . ." he pondered lengthily. But he lost himself in a dream. "Well, whatever."

Teri, wearing a long purple-dotted dress, will be 20 next month. She works in a leather shop in Berkeley. "It's near 'Frisco," she explained.

"I heard that at Mardi Gras there's a lot of people partying and boogeying," she said, sitting in a passive, watchful posture. "I like it where there's a lot of parties and freaks and long-hairs." She was staying with some friends on Burgundy Street and would be in town for two weeks, she thought.

"What do I do here? Well, I go to bars and drink, talk to boys, pick up guys." She prefers Berkeley to New Orleans. There she has an "old man" who works in a Micaseal paint factory for $4 an hour.

"To be truthful," she said, on further reflection, "all I do all day long is boy-watch. That's all I do. And I drink, although that isn't my normal thing." Here she whispered something to an acquaintance who came by.

"He's freaky," Teri said, indicating a strange-looking, nearly unconscious boy on her right. The cave-dweller, to her left, squirted out some more Spar Var. "That paint eats your brain cells up," Teri said.

"What is this place?" said the cave-dweller abruptly, as though waking from a dream.

"You're at the Process," said bearded Brother Patrick with composure. He was quietly tidying up—lifting a coffee cup from a table. "You're all welcome to stay here," he said, addressing the room as though on impulse, "but we want you to know that this establishment is also a church, and we will be holding our Saturday evening service in about half an hour."

A drowsing Alsatian shepherd dog stirred at his feet. "Kind of a pet," Brother Patrick explained, with a touch of embarrassment at this profane intrusion. Someone asked him if the influx of Mardi Gras mendicants had any influence upon the Process Church of the Final Judgement.

"It helps the Karma aspect," he said, picking up a plate. "We call it the Universal Law. What you give you shall receive. We try to provide a lot of community services."

Teri went on quietly observing the scene around her while Steve heaved and breathed. Outside the rain came down in tropical Louisiana torrents.

The sun was out, on Sunday afternoon, as the street people frolicked in Jackson Square. Stripped-to-the-waist, grubby-jeaned figures squirted leather flagons of wine into each others' mouths. Occasionally, choosing as arbitrarily as a lion selecting a victim from a herd of herbivores, the police moved in and took out the occasional cavorter.

Two-week-old dogs (dusty little balls of fur) abounded. Hippies read comics. Others leaped about in cloggy boots, jean jackets, and pants with elaborate patches darned on with much labor.

"Are you havin' a good time?"

"It's the only time to have."

"Last night when I was arrested I had a couple of roaches in my pockets."

"Hey man," said one girl to another, leafing her fingers through her long hair, "how would you like to come to the country with us?"

"Oh, wow, I've got a lot of country to go to. . . . It's unbee-lievable."

A straightly upright family of five, including two tots in ankle-length granny dresses, came into the park. They set to work handing out copies of The Morning Sun, a Jesus paper. Two "straight," plainly dressed girls, one fat and one thin, sat together on a bench and they accepted copies of this paper, as though they were sitting in a church pew.

Without any warning, a black-clad harridan broke loose from a grubby group nearby and approached the two innocents. She wore calf-length boots, and her alluring shape was wrapped about with studded leather belts and sundry straps. A thin ring pierced the left nostril of her dirty face. With a sweep of her arm she seized the Jesus paper from the fat girl's lap and tore it into strips. Then she balled up the strips and stuffed it down the passive girl's blouse.

"You don't need that shit!" she screamed, repeating the phrase two or three times in a wild, unprovoked rage. The fat girl sat motionless: bewitched. Likewise her thin companion. Bad deed done, the harpie swirled away helter skelter, as though acting out a nightmare.

Someone (a walk-on part from an unrelated play) came by with an armload of Mardi Gras beads strung in Hong Kong.

"Somebody layin' out a bunch of beads?"

"There's just been a parade."

"Oh, wow."

Jeannie is a member of the STP family. She always wears the same long red granny dress, and has intelligent features which tend to adopt an expression of mild hauteur. About 50 members of the STP family are in New Orleans this year. "We're having a family reunion here," she explained to an acquaintance, sitting in a cafe where coffee refills are free and sugar is plentiful. She poured enough sugar into her coffee to fill up about a quarter of the cup, lit up a Camel and talked a bit about the STP gang, which has several hundred members nationwide, she said.

"To look at us we're usually dirty," she said, her appearance confirming her statement. She relished this matter-of-fact analysis of her gang's disdain for the normal. "Usually our clothes are the same clothes we've been wearing every day for a year or two. That's very important. Those clothes are what we call our 'originals.' Every time we get a tear we patch it. We never throw our clothes away, it doesn't matter how dirty they get or how much they smell.

"Patches are almost always leather, and we wear blue jeans, bandanas, and conchos sowed onto clothes with leather—which you wear with your jibber-jabber. That refers to useless toys and trinkets, as in coins with holes in them and animal claws, animal teeth, bird wings, or even petrified pieces of shit from much-loved dogs; jewelry, usually not bells; and we wear boots mostly because we hitch-hike from town to town."

She asked for a coffee refill, topped it up to the brim with sugar, and explained that if possible the STP family lives in rented housing in the cities they descend on. Currently, five or ten members of the loosely-knit clan are in jail in New Orleans, Jeannie said. Their main stomping ground is Boulder, which is where the group was founded by a trio named STP John, Bishop, and Little Brother. STP is the name of a hallucinogenic chemical, said to be more powerful than LSD. According to legend, the drug was taken regularly by the family's founders, all three of whom are now dead. STP is no longer manufactured, and the family sticks pretty much to alcohol.

"That's another way you can distinguish us, we're usually drunk," said Jeannie. "I heard it phrased by a dude named Butler in Colorado that the STP is just a bunch of young drunks trying to take care of one another as best they can."

This sentiment is echoed by a Legal Aid worker, who termed STP members "the skid row winos of the future."

Jeannie, 19, is the daughter of a college professor in Nashville. She has been traveling around the country with the group for about a year. She explained how the three founders died. STP John was shot in Boulder—why or by whom she did not say. Bishop, formerly her "old man," "took his life with MDA, a psychedelic narcotic sort of like floor sweepings." Little Brother died of a heroin overdose last December. The latest STP boss is said to be a stocky fellow called Chipper, recently released after a brief jail term here.

By 11:30 at night a steady flow of indigent street people began showing up at Abbitte Inn, 1111 Governor Nicholls Street, a "crash pad" run by the ABBA Foundation. The admission fee is a dollar a night. People with no money are usually admitted anyway. The front room of the hostel had little more in it than a naked light bulb. Brother Ron from the Process sat at a table by the door collecting money and writing down names.

Signs on the wall read: "We Discriminate. This House Is Sexually Segregated. Please Don't Ask." "Lug. Valuables May Be Checked With Management for Safekeeping." "This House Is Not Run Democratically. Dig It." Upstairs were the bare sleeping rooms: separate for males and females as advertised. (But Brother Ron says that for every female, six males spend the night at the Inn.)

Two nattily attired guys from North Carolina showed up, explaining that they had been turned out of the Marriott Hotel on suspicion of being snipers. They wanted to stay at the Inn because they feared that the police would come after them if they tried to check in at another hotel.

Brother Ron explained that he had had trouble from STP members the night before last and that he wasn't going to let them stay if they showed up tonight. A quiet youngster from Mississippi came in and sat against the wall, listening. The next arrivals were Dan, 21, and Doug, 19, both with sun-tanned, open-air looks. They were carrying backpacks. They explained that they had left Portland, Maine, seven weeks earlier. Dan had been traveling around the country for three years, he said, and they had just arrived for Mardi Gras.

Lo and behold! Three STP members came in: Jeannie (carrying two small and apparently distemper-ridden dogs), Tom and Muskrat. They looked like villains swinging into a saloon in a Western. Muskrat was short and dark, Tom tall and blond. They immediately got into a shrill back-and-forth with Brother Ron when he told them he intended to keep them out.

"You're nothin' but a hippie!" Tom screamed. "Capitalist!" Jointly they screamed and shoved and pushed at Brother Ron. His assistant Donna ran to get help from across the street. She returned with Brother Patrick.

"Come on! Let us stay here," implored Jeannie, who was tired and frustrated. "You're nothing but a bunch of hippies!" She had earlier explained that the difference between a hippie and a street person is that a hippie grows his hair long as a stylish affectation; the street person looks the way he does without thinking about it. Still, it is doubtful that the Brothers Ron and Patrick took "hippie" as the great insult it was intended to be.

"Seven years I've been in the street," said grubby Tom. He continued to remonstrate with flaxen-hippie-headed Brother Ron. Inside, and out on the porch, their argument went on for half an hour or more. In the end, the STP trio was turned away. They came to rest, after a wearisome walk, on the crowded floor of the STP flophouse on Chartres Street by Elysian Fields. Here loud shouting and endless bickering kept everyone awake until 5 a.m., when there was a brief lull until the dawn chorus started up once more.

Lonnie, an ABBA Foundation worker with a jagged scar across his left cheek, came into the Abbitte Inn as the dispute was ending. He dis-agreed with Ron's position.

"We are trying to provide places for people who are not so cool that they can stay wherever they want and don't need any help," he said. "The people we must help are the social outcasts who do not have anywhere else to go. That's what we're here for. Where are those people going to stay? If they walk around the streets at night they are bound to be picked up by the police. Going to jail isn't going to do them or anybody else any good."

Brother Ron agreed, but said "people like that are going to cause the place to be closed down because they create so much disturbance in the neighborhood. Plus, they keep everyone else here awake at night."

Dan from Maine, with friendly visage, sat in a corner and discussed the matter philosophically, bringing to bear the experience of his three years on the road.

"If you've lived on the street for seven years and you don't get any breaks," he said, "you're going to act the way that person did, screaming and shouting like that. It can be very tough. If you're doing a lot of drugs, not eating a lot, and then you start hustling, stealing, giving blood, drinking, and on top of that you don't get any breaks, then that is what is going to become of you.

"I'll give you an example," he said. "Doug and I had a break on our way down. Two chicks picked us up in Maryland, took us to their home, so we stayed there for about two weeks. We worked for a week on construction, made a hundred dollars each, and we were able to pay them back the money they had laid out for us. That's the kind of break you can get.

"Also, I'm lucky—privileged. I can make phone calls and charge them to my mother's phone, or even have money sent out to me. So you see people like me are really burdening society—as I'm only just beginning to realize. Places like this are really needed for people who have much more need than me.

"I'm going to go back to Maine soon and get a job," he said. "We'll probably leave New Orleans soon. It doesn't seem worth staying for Mardi Gras day. Also, I'm getting tired of going places without money. Mentally, physically, and spiritually, you get ready to put down roots. So I plan to work for a year in Maine. Then maybe I can save up money to go to Europe. After a certain time on the road you realize that you've had enough, and I realized it watching those people tonight."

He and Doug picked up their backpacks and trudged wearily upstairs for a night of rest on the floor. It had been an exhausting day for almost everyone.

By Tuesday, one week before Mardi Gras, Dan and Doug had left the city, as had the pair from North Carolina suspected of terrorism. Dan from

Lanham, Maryland, was still recumbent on the Jackson Square turf, but wider awake now; Harlan and Debbie were nowhere to be seen. Teri from Berkeley was lamenting the departure of her most recent old man— "split for Virginia Beach." When he was around she didn't particularly love him, was her impression, but now that he was gone she thought maybe she did after all.

STP members Tom and Muskrat were to be seen rampaging about Jackson Square with bottles of Mad Dog wine in hand. Jeannie was running not far behind them. Tom was her new old man, she had decided.

As for Steve the cave-dweller, he was stationed patiently on a Decatur Street corner, smiling at passers-by but warily refraining from asking strangers for money. He knew the police were likely to be down on him in an instant. Tucked into his belt was a copy of Thucydides' *History of the Peloponnesian Wars*, but his stomach was empty. Man cannot live by Great Books, nor by Spar Var spray paint alone.

An acquaintance walking by realized Steve's predicament. He fished into his pocket and handed the long-suffering cave-dweller a dollar bill.

"Outasight!" Steve said, looking down at the money as though it were a long-lost heirloom.

First one, and then a second street person appeared by Steve's side. All three looked down at the dollar bill in wonderment—as though the stories they had heard about Mardi Gras in New Orleans were at last coming true before their very eyes.

Was Sirhan Sirhan on the Grassy Knoll?

My most vivid recollection of Jones Harris is that he always wore a straw hat. Even indoors he preferred to keep it on. Another characteristic was that he never would put anything down on paper. On that score, I remember Jim Garrison, the district attorney of Orleans Parish and my boss at the time, saying that we didn't even have a sample of Jones's handwriting. This was meant as a sort of "black" prosecutor's joke, as though all good Machiavellians prudently stored up evidence against their most loyal friends.

This disinclination on Jones's part to write anything down was most uncharacteristic of conspiratorialists (a deliberately convoluted word I have coined), most of whom were and remain prolific memo and letter writers. And working for Jim Garrison, as I then was, on his ill-fated Kennedy assassination investigation, I met most of the conspiratorialists of those years. Jones Harris is one who remains vividly in my mind.

That was in 1967 and 1968. Now, many years later, I have been looking over some of the books and articles that came out at that time, dealing with the assassinations of the 'sixties, and it has brought to mind my strange experiences with the Jolly Green Giant, as Jim Garrison was known to friend and foe alike. Perhaps by recalling some of them I can shed some light on that lurid time.

First, the reader might be curious to know how I ever became involved in such an outlandish adventure. It all started for me in the fall of 1966. I was almost completely out of money and I found myself teaching at a small, not-very brilliant secretarial school in Baton Rouge, Louisiana. I seem to have developed a mental block about that undistinguished episode in my career and so I won't even try to say anything further about it. The general idea seemed to be to teach these high school graduates reading, writing and arithmetic, and I do recall that there were other courses in typing, office machinery and so on. Baton Rouge, over which the aroma of many oil refineries always seemed to hang, was not the most exciting place, and in my boredom I began reading the books then coming out about the Kennedy assassination, by such authors as Mark Lane and Edward Jay Epstein.

I would return to New Orleans at weekends—blessed relief!—and there I knew a girl called Rae who had befriended an assistant D.A. working for Jim Garrison. His name was John Volz—later he became the U.S. Attorney for the region. Rae introduced me to Volz at the District Attorney's Office at Tulane and Broad (attached to it was a nightmarish prison seemingly filled with black male inmates exclusively). Volz to my surprise took me immediately down a long corridor, to a comfortable panelled office in which the enormous figure of Jim Garrison (6'8" I believe) was sprawled out on a sofa. I took to him immediately. A man of great charm, he regaled me with stories about the poor investigative job done by the FBI. He signed me up on the spot as one of his investigators, without, as I recall, asking me a single question about myself. He turned out to be an irresponsible rogue but I always liked him—particularly for the casual way he hired me. My assignment was to go to Dallas right away to "find out what you can."

Three years after the event the trail was cold, but it sure beat secretarial school in Baton Rouge.

Garrison had confidently told me that he knew who the conspirators were, even seeming to know the details of getaway planes and so on. It was fantastic to think that he really did know these things (talk about the story of the century!) but he seemed so confident and—an important point that badly misled me and many others who were to come after me—was he not an *elected official* with a job that carried with it major responsibility: the power of arrest and prosecution? Was it likely that such a person would simply fabricate such tales? Or would *make-believe* that he knew such vital matters of state, sure to attract worldwide headlines? Not all that likely, I reassured myself, and therein lay my error.

As I made my way to Dallas I felt like a character out of a Raymond Chandler novel . . . except. . . ! Was I not about to solve the crime of the century? My "contact" in the D.A.'s office was Garrison's laconic chief investigator, Louis Ivon, a member of the New Orleans Police Department. I was supposed to "check in" with him every day. His problem was to point me in certain investigative directions, without necessarily letting me know what it was that Jim Garrison was really interested in. Ivon, rightly, no more trusted his investigators (knowing as he did how casually they had been hired) than his investigatees.

As far as I was concerned, Ivon solved this problem by giving me virtually no instructions at all, and politely feigning interest in the odd, useless snippets of information that I occasionally sent his way. It turned out that the FBI, far from having done a poor job, had very thoroughly interviewed anyone and everyone with the most tangential relationship to the assassination, and within about a week of the event. The reports of these interviews were mostly available for public inspection in the National Archives Building in Washington, D.C., and after about a month in Dallas I was able to persude Ivon to let me pursue my investigations in the rather more comfortable surroundings of the National Archives reading room on Pennsylvania Avenue.

But to show you the way conspiracy theorists work, here is an example of something that Louis Ivon did ask me to "check out." (Not that Ivon himself had thought this a promising line of inquiry.) The conspiratorialist art is to forge links and connections between people they seek to enmesh in conspiracy. Indeed, linking people together is the very essence of the business. Garrison himself was in some ways not a typical conspiratorialist: he was too frivolous and sometimes had a most uncharacteristic sense of humor about his theories. But in other ways he was a conspiratorialist in the grand manner—one of the best linker-upper's in the business. Name two people in New Orleans and Garrison could link them. If he had a rival in this regard it could only have been Mrs. Mae Brussell of Carmel, California, author of the "Conspiracy Newsletter," and one of the more delightful visitors to the D.A.'s office in New Orleans. But more about her in a minute.

In assassination circles, one of the main linkage factors is what Garrison loved to call "propinquity." If two people live near one another, say within two or three blocks, it's suspicious. If any closer—they are "linked." If, on the other hand, they live at opposite ends of the city, get a list of friends of each (from their address books). Two such friends are very likely to live in the same block, or even know each other. Presto—the link.

City directories are indispensable tools for conspiratorialists. Garrison would spend hours poring over the New Orleans phone book and "criss-cross," as totally absorbed as another might be by a detective novel. One day he looked up at me from his city directory and said, "sooner or later, because people are lazy, you catch them out on propinquity." He wrote two memos entitled "Time and Propinquity: Factors in Phase Two," which are, I should think, classics of their kind.

Well, imagine Jim Garrison's excitement when he found the following: In Dallas at the time of the assassination there lived a Russian-emigre oil geologist named George De Mohrenschildt who had befriended Lee Harvey Oswald after Lee returned from the Soviet Union in 1962 (whither he had defected in 1959). There was another member of the Dallas emigre community named George Bouhe, who knew De Mohrenschildt (who knew Oswald). And, city directories showed, Bouhe once lived right opposite . . . Jack Ruby! (He shot Oswald, just in case you had forgotten.) And there you have the long-sought Oswald-Ruby link—based on propinquity.

Would I "check out" this link, Louis Ivon asked me one day? (He kept a perfectly straight face, or at least a straight voice, on these occasions.) The boss was interested, he said. Sure, I replied. So I made my way to the Bouhe address—and there sure enough was the house. Opposite stood the Ruby abode. Both houses, I sensed, were guarding their secrets closely. . . . No, seriously, I wasn't quite sure how to proceed. Ruby was dead by this time (suspicious circumstances, need I say), and Bouhe might have been too, for all I knew. I wondered if I should perhaps crouch in the laurel bushes and try to peer into the Bouhe dining room, or what. Nothing in my educational background had prepared me for this predicament. Eventually I decided unilaterally to give up on that particular link. I phoned New Orleans collect.

"Sam Spade here," I said. In the office I was known by this inappropriate code name—one of Garrison's little jokes, I think. (I promise I'm not making this up.)

"Uh-huh," said Ivon.

"Could you tell the boss that I checked out that lead?"

"Uh-huh."

"There's nothing to it, Louis."

"Uh-huh."

Later on I heard Garrison say, in the course of one of his fantastic monologues at the New Orleans Athletic Club, that the phrase "checked out" was one of the cliches of contemporary speech that set his teeth on edge. All too often for him it must have been a prelude to disappointment.

Often, too, when Louis Ivon brought him the news that some far-fetched lead had been checked out, Garrison would have forgotten all about it anyway and moved on to some other area of interest—for example the "involvement" of H. L. Hunt, the Dallas billionaire, code named "Harry Blue" (for true-blue conservative). Garrison incurred the disapproval of the mainstream media, but he was about five years ahead of his time in recognizing that the CIA, the FBI and "true-blue patriots" were safe targets.

After calling in, my day's work done, I would often go and see Mr. and Mrs. Holland, my ultra-conservative friends in Dallas. At one time they had actually been involved with some paramilitary outfit called the Minutemen. But they were very kind to me and I liked their company. Usually, when I arrived, Mary Holland would be in the back room—studying city directories. She was a phenomenal researcher and later made occasional trips to New Orleans, incognito, where she gave Garrison some help, or tried to.

This is worth mentioning because it brings out the point that normal ideological labels do not apply to true conspiracy believers. They are neither on the right nor the left, but somehow united on some other dimension. Jim Garrison was in many respects a liberal. He used to pride himself, for example, on his refusal to discern a Communist conspiracy in the Kennedy assassination, and he loved to deride J. Edgar Hoover's "obsession" with Communism, saying that Communists were like unicorns: everyone had heard of them but no one had ever seen one. Nonetheless, he was always on good terms with conspiracy-minded right-wingers, and they with him. When conspiracy discussions were underway, left-right distinctions were unimportant compared with the joy of discovering linkages, connections and overlaps. Moreover, all were prepared to agree that the spectral "they" who controlled the nation but who were so rarely identified were equally inimical to left and right.

I was about to say something about Mae Brussell. I believe that she actually had D.A.-investigator credentials for a while. I know that the West Coast satirist Mort Sahl did. What an odd bunch we were, in retrospect! Although I must say that I always liked Sahl. His interest in the investigation was perfectly sincere, and he could be tremendously funny (about almost any subject *except* the investigation). Years later I used to hear him hosting a phone-in radio show in Washington D.C. He used up a good deal of his air-time having fun at the expense of liberals.

Mae Brussell reminded me somehow of Margaret Rutherford playing amateur detective in a British movie. By the time I met her I had moved back to New Orleans from the National Archives in Washington and had a small office where I was in charge of putting in order the numerous files we were beginning to accumulate. Mrs. Brussell arrived from her Carmel retreat loaded down with weighty parcels and packages of documents precariously knotted together with string. She proceeded to unfold a linkage chart, multi-colored, that more than covered my entire desktop. It linked, as far as I could see, everyone to everyone else.

Jones Harris of the straw hat frequently used to arrive in New Orleans with his good friend Richard Popkin, a professor of philosophy from San Diego who was on good terms with Bob Silvers of *The New York Review of Books*. Jones was the son of Jed Harris, a well known Broadway producer, and the late actress Ruth Gordon. He seemed to know everyone you had ever heard of in New York. It was through him that, in swift succession I had lunch with Bob Silvers, dinner with Norman Mailer, and afternoon tea with Lady Jean Campbell, Lord Beaverbrook's granddaughter and Norman Mailer's third wife. Later Jones married a Vanderbilt, and at the reception described himself to a *New York Times* reporter as a "freelance researcher." That is exactly what he was when I knew him.

Jones's friend Richard Popkin was the author of an implausible book called *The Second Oswald*, and it was on this issue that I began to realize that I was wasting my time and would eventually have to break with Garrison. In the course of its investigation, the Warren Commission produced some evidence that Lee Harvey Oswald was in two places at once—manifestly impossible. Just about all of this evidence derived either from fleeting observation or from unreliable witnesses, and so could be dismissed as mistaken identity. In the course of so wide-ranging and well-publicized a case it would be surprising if some such evidence did not emerge. Nonetheless, Popkin chose to postulate two Oswalds.

I had my doubts about this. Imagine planning a conspiracy and starting out by finding two people who look alike—with a view to laying a confusing trail and subsequently misleading investigators. It is a very implausible scenario. So much easier to conclude that a small handfall of people only *thought* they had seen Oswald (at a place where he could not very easily have been).

One day I had dinner with Jones at the Ponchartrain Hotel on St. Charles Avenue in New Orleans and I outlined this and other difficulties with the "second Oswald" theory. There could hardly have been two Oswalds, I said. Jones replied, perfectly seriously: "You're right, Tom. There were three Oswalds."

I gaped, temporarily forgetting my Creole gumbo.

"And two Rubys," he added.

About that time, I began to give serious consideration to the possibility that Lee Harvey Oswald shot the President alone and unaided.

More recently I heard from a reliable source that Jones was still at work on assassination theories. He was reported to be wrestling with a sort of unified field theory, linking most of the assassinations of recent years. Such a theory would of course not only represent a great breakthrough but would greatly simplify life. Instead of four loners we would have one mastermind. One suspects, however, that just as Einstein discovered insurmountable obstacles in his efforts to unify the theories of gravity and electromagnetism, so the difficulties confronting Jones Harris will prove in the end insuperable.

Not long after Jones outlined his "full house" of Rubies and Oswalds, he took me to P. J. Clarke's in New York, where the big event was lunch with Bob Silvers of *The New York Review of Books*. I should explain that I was not entirely looking forward to this encounter, because by this time Professor Popkin had published in Silvers's journal a second, and I thought untenable article, which tried to quell the by-then raging press criticism of Garrison, on the grounds that Garrison had not yet had his day in court. This view embodied an illiberal theory of justice not normally to be found in *The New York Review of Books*. Did one speak of the *prosecutor's* "day in court"?

I was therefore worried about what to say if, as I suspected, Silvers at some point during lunch leaned across to me and said, in a "between-you-and-me" sort of way: Does Garrison really have anything, or not? By that time, Clay Shaw, a prominent New Orleans businessman, had been charged, along with a strange character called David Ferrie, with conspiracy to assassinate President Kennedy. (Ferrie died of a cerebral hemorrhage a couple of days after the charges were filed, much to the relief of one or two senior attorneys in the D.A.'s office, who knew that the case against him was weak to non-existent.) But as it turned out my fears were groundless. Silvers was a perfect gentleman and didn't ask me any embarrassing questions about the case. The lunch went off smoothly, with the whole subject of the assassination—perhaps on the "don't talk shop" principle—rather pointedly ignored.

I had been at the National Archives when the charges against Shaw were filed, and I managed to stay on there for a few months longer, sharing a house on Ashmead Place with two British journalists who became and remain good friends, John Graham, then with *The Financial Times*, and Dominic Harrod, then with *The Daily Telegraph*. But when I returned to New Orleans it became clearer and clearer, much to my disappointment, that Garrison's whole case was based on the testimony of witnesses who were unreliable to put it mildly. I knew that I should do something—what, exactly, I wasn't sure—but I was enjoying myself so much that one way or another I contrived to procrastinate, meanwhile remaining in Garrison's not very arduous employ.

I learned a lot about America in that time. In particular I became acquainted with many members of the press corps, who on the whole did a good job in the Garrison case, and whose influence I believe really did constitute a textbook "check" against the arbitrary exercise of government power. It wasn't until several years later, after Watergate, that this mounting capacity to embarrass and undermine public figures was taken so far as to threaten to create an entirely new and unexpected danger— one posed by the news media itself. It is worth noting also that in 1967-68 the press corps was less ideologized than it would become. Garrison's charges that the CIA and the FBI were "involved" either in a cover-up or in the actual assassination impressed few in the media outside *Ramparts*, *Playboy* and the *Village Voice*. Ten years later, when similar and equally unfounded talk emerged from a House Select Committee, *The New York Times* took it much more seriously.

Not long after I met Bob Silvers Jones Harris introduced me to Lady Jean Campbell, then a New York correspondent for *The London Evening Standard*. She was a sensible lady, having become involved in Jones's odd activities, as far as I could see, partly out of a desire for a "story," and partly out of a sense of adventure. (Possibly also boredom with the East Side life. By that time Norman Mailer had moved on to his next wife.) Much earlier, shortly after the assassination itself, she had traveled on investigative expeditions with Jones to Dallas.

One day Lady Jean came down to Washington, where I was then staying, and we went off together on our rather hilarious trip to West Virginia, in fruitless search of Kennedy's killers. The background was this:

By this Jim Garrison time had irrevocably succumbed to the temptations of publicity. Those who knew him well say he was at first stunned, then soon addicted to the huge, push-come-to-shove press conferences, the camera crews on the double, the blinding television lights, the stream of visitors, the journalists who paid court (for some did)—in short, the opportunity to perform.

But after a time the media went home and Garrison missed the attention. He was given the wrong kind of encouragement by a very intense conspiratorialist named Vincent Salandria, a Philadelphia lawyer who believed passionately that "they" were on the verge of somehow taking over the country. Once Salandria came to New Orleans and in front of the entire assembled D.A.'s staff urged Garrison to make more and more arrests. He saw this as the only feasible way of stopping "them" from proceeding with their dastardly plans. Garrison, of course, saw it as just the advice he had been looking for.

Shortly thereafter Garrison did in fact arrest one Edgar Eugene Bradley of Los Angeles, charging him with the same crime as Clay Shaw—conspiracy to assassinate the President. It was never entirely clear to us in the D.A.'s office whether this was the same conspiracy as Shaw's or another one. Had two entirely separate gangs opened fire simultaneously in Dealey Plaza? (Don't laugh, lest someone tell you there were three.) Bradley was actually charged on the basis of a letter sent to the D.A.'s office by someone who bore Bradley a grudge. Later the charges against Bradley were dropped and I heard that he and Garrison actually became friends. Garrison would even drop by to see him on his frequent trips to the West Coast. This may seem an unlikely turn of events but knowing Garrison I can believe it.

Anyway, the press took little notice of the Bradley arrest, which miffed Garrison a bit. I'm sure he was hoping I would bag him another conspirator when he dispatched me to West Virginia to interview, and above all to photograph, one Jack Lawrence, who had been in Dallas on the day of the assassination. Lawrence had worked in a Lincoln Continental dealership where the "second Oswald" had put in one of his ghostly appearances.

Lawrence had left work early on the morning of the assassination—he told the FBI that he had gone home to rest—and had even parked his car somewhere in the area of the now infamous Grassy Knoll bordering Dealey Plaza. Some of the salesmen at the automobile dealership had told me they thought Lawrence had mentioned something about having been in New Orleans at some point. The plan, therefore, was to get pictures of Lawrence to show to our witnesses in New Orleans—some of whom were remarkably adept at making identifications, especially if they had narcotics or other charges hanging over their heads. Then, who knows, a third conspiracy to assassinate the President could well be in the works.

Before leaving D.C., I rented a camera—one of those intimidating affairs with a huge telescopic lens. My knowledge of how to use it was vague at best, and as I fiddled about trying to make it work, I remember thinking how sorely I lacked the expertise of those "agency" boys over at Langley whom Garrison was always expatiating on, and who no doubt had me in their surveillance at that very moment and were laughing their heads off.

Lady Jean arrived and we drove off to the mountains in a rented car. She had sensibly brought along an Instamatic camera. Her plan was for us simply to knock on Lawrence's door and pose as reporters writing a book about the assassination. After hours of driving through the mountains in a snowstorm we reached Charleston. Lawrence was quite willing to talk to us, and as he did so his wife dandled a baby in her lap. It soon became clear to Jean Campbell and myself that he was either perfectly innocent or a very good actor. (Not that the involvement of actors in previous assassinations is unheard of!)

Then came the irony that I have long savored. Lawrence told us that *he* was beginning to suspect a conspiracy. (He brought up the subject, not us.) He even took us into a back room where he showed us a shelf of "true blue" paperbacks, of the type that are put out by ultra-conservative publishers, proving to his satisfaction that there was some kind of leftist conspiracy in the United States. I couldn't help reflecting, however, that in his case his ideas were not entirely paranoid. There really was a reckless D.A. a thousand miles away contemplating arresting him! And we were a part of the conspiracy. . . .

Lawrence lived in a very modest house, and life cannot have been easy for him. I believe that he worked for one of the coal or steel companies. It struck me as odd at the time that someone of his social status saw the country as beset by a conspiracy of the left. No longer does it seem so odd.

He permitted Lady Jean to snap a picture, which came out fine and was happily not "identified" back in New Orleans. Garrison, I think, was not entirely pleased with my negative report. But his staff, getting fed up with filing charges against unknown individuals in various parts of the country, was delighted. Sometimes I think that this trip to West Virginia was one of the few useful things I accomplished during my period with Garrison.

I have mixed feelings about this, but I am unhappy to report that I ended up betraying Garrison to Clay Shaw's lawyers. I feel partly justified, in that Garrison was blithely wrecking the reputation of an innocent man; but at the same time I feel ashamed to admit it, because Garrison had always treated me well, in some respects almost as a "court favorite," regularly including me in his soirees at the Royal Orleans Hotel and other French Quarter spots.

There is no "discovery law" in Louisiana, which means that the prosecution can produce surprise witnesses at the trial. In the Clay Shaw trial Garrison was preparing to put on the stand a witness who, apart from testifying that he had heard Shaw discussing assassinating Kennedy, was under the impression that the government broke into his house from time to time and substituted "dead ringers" for his children, who then spied on him. (To foil the fiends, he periodically fingerprinted his children.) He believed that he had been hypnotized at least 50 times against his will, and so on. But the truly confusing thing was that he was perfectly normal in most respects, and held a good job as a certified public accountant. On his return to New Orleans, the assistant D.A. who had interviewed him in New York said, standing next to me, "Well, he'd make a hell of a witness, but he's crazy."

For a while I assumed that Garrison wouldn't use him as a witness. When I found out that they really were going to, I decided that the time had come to forewarn Shaw's lawyers, thereby giving them time to look into the man's background. The man did testify (indicating better than anything the shortcomings of the case against Shaw), and the investigative material on the man was flown in from New York, arriving in New Orleans barely before the cross-examination began. For my perfidy I was charged by Garrison with "unauthorized use of movable goods," a misdemeanor charge that was not brought to trial and was later dropped.

A couple of final memories. I remember once having dinner with Garrison and a few others at a French Quarter restaurant in 1967. Surprisingly, his mother was with us at the table. A large and imposing woman, she sat at the opposite end of the table from her son. Mark Lane was also present.

"What if Walter Jenkins were involved?" Garrison said at one point, referring to the LBJ aide who had been charged with disorderly conduct at the Washington, D.C., YMCA. "Perhaps I should have him arrested. Can you imagine the headlines . . . " (holding up his hands to show the space they would fill.)

"Oh Jim, I think that's a wonderful idea," his mother said.

By the end of the evening Garrison seemed convinced of Jenkins's guilt. But, typically, the next day he had forgotten all about it. By then he was playing about with another scenario, this one including Allen Dulles of the CIA, who was thought certainly to be "involved." Garrison's plan that day was to have Dulles arrested along with Gordon Novel, a former Garrison employee who had absconded with some potentially damaging information. Both Novel and Dulles smoked pipes, and Garrison was laughing at the prospect of seeing their photos side by side in the papers—both solemnly smoking pipes.

After a day or two of this, Garrison would usually retire to the sauna room of the New Orleans Athletic Club, sometimes for several days at a stretch. After one such recuperation he returned to my office with a one-act play he had written about a mad king who held court wearing roller skates. I kept it in my big filing cabinet, along with the files on Lee Harvey Oswald, Clay Shaw, Robert Kennedy (a suspect until his death), H. L. Hunt, Paramilitary Organizations, Jack Lawrence, the Minutemen, LBJ (clearly involved), Edgar Eugene Bradley, and many others whose names I have now forgotten.

Later on Jim Garrison was charged with violating Federal anti-racketeering statutes. According to the indictment, his former chief investigator, Pershing Gervais, was wearing a small tape recorder on his person as he paid Garrison weekly bribes from organized crime figures to keep their operations free of police interference. Garrison displayed great originality in responding to the charge. At one point he had himself indicted by the Orleans Parish Grand Jury and charged with the state version of the Federal crime. His plan was to win a speedy acquittal in a compliant state court and then claim "double jeopardy" when the Federal trial began. It raised eyebrows even in Louisiana; Garrison never followed through on it. But in Federal Court he shrewdly fired his own lawyer in mid-trial and thereby had the opportunity to act as counsel and defendant simultaneously. There was nothing to stop him from making speeches to the jury, and when he did so he was in top form.

"I tried to find out who killed your President," he said, "and look what happened to me." He was acquitted.

As may be imagined, I fell into disfavor with Jones Harris and some of the other conspiratorialists. I heard later that I was presumed to have been a CIA agent all along. If Jones reads this book, he will learn for the first time of my 1961 encounter, at the Wigmore Hall in London, with the CIA's Henry Pleasants. One more link—and rather a suspicious one he may well think. I wouldn't blame him. One of these days I really must file one of those Freedom of Information lawsuits, to find out for myself what's in my CIA file. After my Garrison experiences, I shudder to think.

Oysters at the Pearl

The other day I received a phone call from the Washington correspon-dent of the London *Financial Times*, advising me that one of the paper's European correspondents would be coming through New Orleans in a couple of weeks. Could I show the fellow around and perhaps give him one or two ideas? He was new to the South apparently.

Happy to, I said, and then sat down to try to think up a new wrinkle on the "New South" theme. I realized this was getting a little iffy from the *FT's* point of view. There had been at least three stories on the subject in the paper. One said how much things had changed, another said how little they had changed—both written by me. Then the Washington editor himself put in an appearance and deftly produced a third story compatible with the first two—quite a feat. Now a fourth was needed, apparently. I worried that the remaining interpretations left open to us were becoming distinctly limited. Only so much of this insipid "New South" stuff could be published, surely. Moreover, the memory of Bull O'Connor and civilly righteous marchers was growing just faint enough to make conscientious editors worry: Was there really enough of a story in the South to justify all that expense money?

Travel, of course, is one of the unstated goals of a career in journalism. At any rate, I was fairly sure that this was the real purpose of the new man's visit. Number Four, I shall call him. My suspicions were height-ened by the news that his girlfriend would be coming along with him.

In due course Number Four arrived. I met up with him and his girlfriend at the Pearl, a well-known lunch place downtown. (She was already beginning to suffer from the unexpectedly rich Creole cooking.) After we settled down at a table with oysters on the half shell and misty-cold mugs of beer, I outlined my new "angle" on the South.

"It's very simple," I said, squeezing some lemon onto an oyster. "There *is* no South any more. It's in the past, gone with the wind, retired, demobilized—there's your story right there."

Number Four looked disappointed, I could tell. His beer mug halted in its upward journey to his lips, and was replaced on the table. His girlfriend frowned at me.

"Let me explain," I said, eager to maintain my credentials as a worthy "stringer" for the paper. "First of all, 'the South' is an area about the size of Western Europe. Alabama alone is bigger than England. Other than membership in the United States, what can so huge an area possibly have in common? Of course in the 19th century there was something—slavery. Then in the hundred years following abolition there was still something—the South's reluctance to accept the result of the Civil War, assisted by such concepts as states' rights, continued disenfranchisement of blacks, and so on. But now all that really is in the past. So, I believe, there's no such thing as the South any more—except in the purely geographical sense, of course."

"Oh, I must say we haven't found that said," Number Four, who had been busy with his oysters, occasionally glancing up at me as I spoke. "The moment we left Washington and drove south, we found we only had to get out of the car and go into some roadside place, and people would immediately bring up the subject of race. We didn't find that anywhere else in America."

Number Four elaborated on this, revealing that his travels through the South had been more extensive than is customary in such cases. He had actually been on the road for a whole month, winding through obsure portions of North Carolina and Georgia and so on. Everywhere it had been the same, apparently. He had already filed several stories, he told me. Plainly he was committed to the tried and true, old-South-is-still-with-us gambit—one that is more than amply explored in such regional journals as *Southern Voices* and any number of national ones based in New York. Number Four wasn't about to change his mind at this stage of the game. New Orleans was his last stop. Soon he would be flying back to New York—then on to his new assignment in Paris.

"Of course," he added, by way of a concession, "I imagine you could find whatever story you wanted if you set out to look for it."

Of course. He may have imagined that he was coming to the South with an open mind. No doubt he really did come with an open mind. But put yourself in the position of those roadside Virginia farmers he no doubt encountered after leaving Washington in his rented car. It's a burning hot day and Number Four is dying of thirst so in he goes to the next roadside store. There's a heavy, creaking door covered with wire netting that he pulls open against a tough spring. Inside it's cool and dark and the door snaps shut behind him.

At first he can't see the farmers standing about but they can see him all right—a sunscorched British beetroot with unbecoming Bermudas three sizes too big coming down to his knees; incongruous black walking shoes and greyish white legs: the last time *they* saw the sun was at Henley-on-Thames. He is perspiring freely and folding a Mobil map.

"I say . . ." he says tentatively. This is the hard part. He has to Do Research at this point to justify the whole enterprise and reassure himself and his girlfriend (who's wandering about outside, keeping an eye on the car and wishing she were *back* at Henley, if the truth be told), that he's not one of those hacks whose stories consist of imaginary quotes dreamed up in the hotel room.

The farmers look at him without saying a word, but they're already thinking that this one reminds them of the one who came through a couple of weeks earlier. And sure enough he tells them that he's with one of those foreign newspapers—something to do with writing about "the South." Looking for trouble, in other words.

Before he came in the door, of course, the farmers were talking about tobacco prices, inflation, and the latest farm regulations out of Washington. But they know this reporter don't want to talk about things like that—shop talk. So they puzzle on it for a minute and finally one of them says: "What do you think Earl? I don't think they're *ready* for it myself. . . ."

The Englishman smiles. "Good copy, good copy," he says to himself.

Meanwhile my "No South" variant has yet to be deployed. I shall wait a decent interval following publication of Number Four's reversion to the "Old South" gambit. And then I think I shall give it a try. It will have the advantage of novelty, but there will be an accompanying editorial headache: What do we do for an encore?

Criminal Neglect

By 1974 I had growing reservations about the liberal orthodoxy, but I tried to push the whole subject out of my mind. I told myself that I had not come to America to think about or argue politics. The subject had always struck me as beneath serious consideration (as it does so many young people). But some things you couldn't help noticing—Watergate, for example. I watched some of the Ervin hearings on television. I took the Legislative side, against the Executive, without any great enthusiasm, but was willing to accept for the sake of argument (or rather for the sake of avoiding arguments), that this new form of theater was a contest between Good and Evil. I enjoyed it, too, in a gossipy kind of way ("Have you heard the latest?" "No, what?" "John Dean just testified that. . . ."); and I couldn't help noticing how hypocritical were the liberal incantations of "crisis." Everyone I knew was enjoying Watergate immensely and eagerly looking forward to the next installment, saying all the while what a grave crisis it was. Nixon had "lied to the American people."

Then came President Nixon's resignation, and I didn't want to think about the subject any more. I assumed everyone else felt the same way. The Watergate Show was over and now we could all go back to work. But the vindictive climate persisted. Every day there seemed to be more rage and ferocity in the media. Now President Ford had pardoned Nixon! One was supposed to become indignant about that—why, I couldn't tell. I just wanted to study the theory of evolution but it was getting harder and harder to stay clear of distractions. No one was admitting it but America was under attack, for what reason it was hard to say. Next it was the CIA's turn—apparently they had been spying on the American people. (About bloody time too, I caught myself thinking, one day. Lord! Was I becoming a right-winger? These things do steal up on one!)

Then there was the energy crisis. One day I went to Baton Rouge to interview the governor of Louisiana, Edwin Edwards, for *New Orleans* magazine. He explained to me that the origin of the crisis was very simple and straightforward: oil and natural gas were held under price controls by Federal edict. A kind of price dictatorship had been imposed by Washington. But it was quite impossible to tell my liberals friends about this—especially journalists. Oh, they would say, the Ford Foundation has the answers on the energy crisis. Call David Freeman—he's got the story. Edwards was in the pocket of the oil companies, they said, and couldn't be trusted.

Then came the Vietnam collapse and the ignominious American departure from Saigon. This was greeted by more liberal cheering and guffawing. By this time I knew they were up to no good, but I couldn't understand what was going on around me. Why did so many educated people seem to detest their own country so heartily? Did they have no appreciation? Did they not realize that liberty was a scarce commodity? What was the matter with a country whose university graduates were so often filled with rancor and self-disgust?

At about this time, toward the end of 1974, I became an American citizen, raising my right hand in the Federal Courthouse on Royal Street in New Orleans. I think that was a turning point. Again, one or two acquaintances expressed something between amazement and amusement that I would *want* to become a citizen of a country they apparently could not abide—a country whose leaders, whose intelligence agencies, whose oil companies . . . I hardly need complete the litany. They seemed to think of patriotism as something to be derided unquestioningly. But by then I was myself one of the American people that had been lied to and spied upon and felt entitled to tell them where to get off. And so began my career as a conservative.

I arrived in Washington, D.C., on Labor Day, 1975, having driven a rickety VW from New Orleans. Charlie Peters had offered me a job at *The Washington Monthly*. Charlie was (and has remained) a member of the liberal church, but as a dissenter within it. Later he became the unofficial guru of the neoliberals. By the time I arrived in his Dickensian little rabbit-warren of offices on Connecticut Avenue, N.W. I already knew I was a conservative. Reporting for duty at the same time was David Ignatius, reputed to have been fiercely radical in his Harvard *Crimson* days. (But we got along fine. He had an objectivity unusual among journalists and his later reporting from the Middle East for *The Wall Street Journal* showed no trace of bias.) I feared ideological clashes with Charlie

Peters, but as things turned out there was no problem. There were certain things he had always wanted to say in his magazine, but had been unable to say because (evidently) his young proteges (and writers) weren't about to discard their liberal credentials. One such project he referred to as his "right-wing crime story." I was happy to do it.

I remember being astounded by what I found out when working on this article. My role in it as the surprised, naive observer was not invented. At the same time I became convinced that there is at the core of contemporary liberalism something insane and self-destructive and (there is no other word for it) wicked. Individual responsibility for freely chosen action is denied. The criminal merely makes manifest the injustice of his environment. I have since avoided crime stories. They upset me too much. Since then so many liberals have been indiscriminately mugged that I gather criminal procedures have tightened up somewhat. But things are not dramatically different today.

"YOUR FIRST ROBBERY IS FREE"

"Can I offer you a ride home?" my host asked at the end of the evening, shortly after I arrived in Washington. But it was a pleasant evening and I felt like walking. I knew about the D.C. crime problem, of course, but this was one of the safest areas of Washington—just off Embassy Row. Besides, wasn't crime gradually diminishing? Weren't we turning the corner, especially in Washington, the home of good intentions?

I might have been less confident had I been reading the crime statistics: up 15 percent in D.C. last year [1974], up 17 percent in the nation, according to the FBI. "Crime is increasing at a record rate," according to *The Washington Post*. About 5,000 serious crimes a month are reported in the nation's capital, which is second only to Cleveland in homicides. And there are 4,586 police officers in the District of Columbia, costing each resident an average of $124 a month—double the national average.

Here's another item from *The Washington Post* I missed before setting off on my evening walk: "Recently revised park police statistics show intense criminal activity on the Monument grounds during the afternoon hours of Human Kindness Day. . . . Park Police reported 468 robberies alone on May 10 [1975] on the Washington Monument grounds—more than half the total of 733 robberies reported by D.C. police throughout the rest of the city for the entire month."

Gallup polls have shown that one household in four was hit by crime in the past year, one American in two is afraid to walk in his neighborhood at night. But I was cheerfully ignorant of all this as I walked along. . . .

Hullo, what's this? I'm walking close to the P Street bridge over Rock Creek. Looks as though there's been an accident: the circling red light of an ambulance, police cars, a sheet on the road. . . .

It was a murder, not an accident, committed perhaps only minutes before I arrived. I read about it two days later in *The Washington Post*:

"A minister's son, who came to Washington in July after his college graduation to work as a draftsman for Amtrak was fatally stabbed Friday night, D.C. police said, as he walked along P Street N.W. near the bridge across Rock Creek Park. Duncan McCrea, 21, died shortly after 11 p.m. of a single stab wound in the chest, according to the D.C. medical examiner's office. Amtrak officials said they believed McCrea was on his way home when the apparently random killing occurred."

A man named Aubrey Dockery was soon arrested and charged with murder. He was arraigned the next day before D.C. Superior Court Judge Theodore Newman. The assistant U.S. Attorney, John Gizzerelli, asked the judge to hold Dockery on $10,000 bond. Instead the judge released the man on his own recognizance—that is, set him free without requiring him to pay any bail—until his trial date.

A few days later I read a *Washington Post* editorial on the P Street murder. The newspaper discerned a "deeply troubling question" arising out of the pretrial release of the suspect, wherein the judge had been "following the law, literally and explicitly." To wit: "Does this procedure adequately protect the community?"

I began to feel that I had fallen into a topsy-turvy world, like Alice in Wonderland.

"The desirability of jailing murder suspects is not so obvious as it might look at first glance," the *Post* cautioned, in its best consciousness-raising manner. "The suspect in the P Street case is far from the only person to be freed here in recent years while facing a trial that could end in a life sentence. The practice is, in fact, not uncommon. Experience has shown murder suspects to be well behaved while awaiting trial. . . . It needs to be repeated that the suspect is presumed innocent until convicted, and being in jail makes it much more difficult for him to present an effective defense at his trial."

The information about "well behaved" murder suspects is misleading, however, because half of all the murder cases the courts deal with involve a marital or domestic flare-up, a "crime of passion," in which it is safe to assume that the attacker does not have a criminal disposition, and is immediately docile following the assault. Also the suggestion that jailed defendants find it difficult to present an effective defense is not plausible. Defense lawyers are entitled to interview their clients in prison.

About a month later I read another and in some ways more remarkable story by Ron Shaffer in the *Post*. A girl named Sally Ann Morris had been shot in Georgetown: "It was about 10:30 p.m., she recalled. She and her boyfriend, Henry Miller, were walking down 33rd Street, heading for an M Street restaurant, when two men approached. As they passed the couple, one of the men pulled out a gun, cocked it and stuck it in Sally Morris' back. 'I knew right away they were going to fire because you just don't cock a gun without a reason,' Morris said.

"Instinctively, Miller grabbed her and they started to run. After a few steps, she said, she heard gunfire and felt a slap at her back. 'It felt like a burning needle that went through me real quick. It sort of numbed me.' The bullet ripped through her intestinal tract and lodged in her lower abdomen. Doctors had to perform a colostomy, rerouting the undamaged intestinal tract to a substitute opening in her lower abdomen."

And there was this: "Police said the two men had been committing armed robberies in Georgetown for several weeks before the Morris shooting, escaping by hiding in the back seat of a getaway car driven by [two] women."

And this unbelievable paragraph: "compounding all this is the fear that the ordeal is not yet over and that her assailants may return to kill her. Four suspects arrested in the case, who were released on personal recognizance, pending trial, promptly disappeared and are at large today." A *Washington Post* editorial soon followed, making three recommendations: "First of all, the District of Columbia needs legislation to compensate the victims. . . ."

First of all? This ordering of priorities disturbed me, not to mention the idea that if only the victims of crime can be paid off, then society and its criminals will once again be in a state of balance—with its familiar suggestion that money can solve all problems, including criminal assault. Would money repair Sally Morris's intestinal tract? What about the release of the suspects? Did that not raise a more urgent issue?

This turned out to be the *Post*'s second recommendation: "The time has come for Congress to hold oversight hearings on the Bail Reform Act and they way in which it is working in practice. Experience under it is beginning to raise questions as to whether the rights of the public are being adequately protected."

I decided to visit the D.C. Superior Court where, I hoped, I would be able to find out more about the Sally Ann Morris case, and the Bail Reform Act itself. The District of Columbia is a Federal jurisdiction, and so it is important to bear in mind that what would be a state crime anywhere else is subject to Federal court procedures in Washington. The prosecutors for local or "street" crime are U.S. Attorneys. Thus the Bail Reform Act of 1965, which applied to Federal jurisdictions, covers *all* crime in Washington.

"I remember the case," said Assistant U.S. Attorney Judith Hetherton. "A girl was quite seriously injured. I argued for a money bond. I take it the suspects didn't show up for the preliminary hearing?"

"Right."

"They were arrested on two separate charges that evening," she went on. "We took our strongest case and held them on that charge. The judge knew they were suspects in the other case, but he has to consider their record of convictions, their ties to the community, their employment record, and so on, in determining their bond. You should talk to Lee Cross. She might be able to tell you more about the case."

Assistant U.S. Attorney Lee Cross, the deputy chief of the Grand Jury section, has worked in the U.S. Attorney's office for four and a half years. The Grand Jury section had rooms where policemen were sitting playing cards, waiting to testify; other rooms with rows of seats bolted together, where waiting witnesses sometimes lay horizontally, giving the appearance of a railway station at 2 a.m.; defense attorneys who strolled from room to room (generally they wore three-piece suits of an elegant cut); and prosecutors with piles of documents under their arms (generally they looked harassed).

Lee Cross sat behind her desk, which was laden with files. She had time to talk, as it turned out, because she was waiting for her 7 p.m. appointment.

"I don't know the details of the case," she said when I mentioned Sally Ann Morris. "But you could look up the 'jacket' in the Bench Warrant section. It's a public document. A warrant will have been issued for the arrest of the defendants. As for the Bail Reform Act, the D.C. Code, Section 1321, discusses release prior to trial. Here's the main point. The judge may not set financial conditions on release merely to ensure the safety of the community. He may only do that if the prosecutor shows likelihood of flight by the accused."

Merely to ensure the safety of the community? Could that be right?

"Here's the wording," she said, showing me the D.C. Code: "*No financial condition may be imposed to assure the safety of any other person or the community.*"

That seemed odd, I thought.

"To hold them, you have to use preventive detention," Lee Cross went on, "and you will recall the screaming that went on among liberals when that was passed by Congress."

I hadn't realized that "preventive detention" meant detaining until trial suspects charged with rape and murder. I thought it had been a repressive scheme to lock up radicals—devised by President Nixon. (Two days later a reporter I knew at National Public Radio told me that was exactly what it was and refused to believe otherwise. When I told her that an Assistant U.S. Attorney had given me the facts she gave me a dark look and told me that I had been duped by the prosecutors.)

"But preventive detention is so hard to apply it's not often used," Lee Cross continued. "The government has to reveal practically its whole case to show probable cause to the judge; the defendant has to be brought to trial within 60 days; he has to have been convicted of a violent crime within the past 10 years; and the government has to show that no other condition will reasonably assure the safety of the community." According to one figure I saw, preventive detention has been used only 50 times since it went into effect in 1971.

"How often do defendants released without bail not show up for trial?" I asked.

"It's disturbingly frequent," she said. "To get exact figures you should talk to Peter Chapin, who is in charge of the Felony Bench Warrant section, and John Hume, chief of the Felony Trial section. If you had more intelligent criminals, a lot fewer would show up, I can tell you that. Fortunately for us, most of them aren't too bright. It isn't easy to track these people. Under the provisions of the Social Security Act, they can go and get a job as a plumber, say in Seattle, use their social security numbers, and the Social Security Administration won't give that information to the FBI, even though they're wanted for armed robbery."

That seemed odd, too.

"Often the defendant is back on the street before the victim leaves the hospital," Cross said. "A defense attorney just told me that two clients told him, 'Your first robbery is free.' Routinely they get probation for armed robbery. God only knows why." She lit a cigarette and considered the question before continuing: "So many judges look at it from a middle class perspective. For a professional person a felony conviction is a serious matter: It's on his record, he loses the right to vote, loses the right to get certain government jobs, various trades won't take convicted felons, you can't practice law, and so on. But for most of the people we get, these considerations don't bother them at all. The *only* thing they worry about is getting time. The rest they could care less about. I had a defense attorney tell me about a case the other day, a client who pleaded guilty to forgery in return for a likely probation sentence. 'I tried to tell the girl what she did was wrong,' the lawyer told me, 'but she thinks I'm crazy. She got 1,500 bucks out of passing bad checks, and she isn't going to jail. That's all she cares about.' She beat the system, in other words."

Lee Cross ran a finger through her hair, glancing aside absently for a second. "I'm convinced that the big unanswered question in all this is witness intimidation," she said. "We have two or three good cases a day dismissed because we can't get the witnesses to come down here. In some cases it's overt intimidation—a phone call from the defendant—but more often just a feeling because they saw the guy back on the street. The Sally Ann Morris case you are referring to? We have much worse ones than that, where the witness and suspect live together on the same block and know one another.

"In such cases as these, the concept of 'innocent until proven guilty' is a pure legal abstraction. You can't expect the victim to believe it when he knows the person who robbed him. Or take the case of the person who is arrested right at the scene of the crime. You can't expect the victim to believe his attacker is innocent until he's been brought to trial."

According to one study, as many as 1,000 cases in D.C. are dropped each year because of witness intimidation. "And here are some statistics we don't have," said Cross. "How many people are deterred from reporting crime because of probation and bail? How many people go to police line-ups and don't identify the person because they know he's already made bail? Many of them refuse to come down here to the Grand Jury and testify. Sometimes they disappear completely. The police can't find them. Or they come down here and say things that just aren't so, like, 'I saw him standing by the truck, but I didn't see him get out of the truck.' Or they'll say: 'No, that isn't the guy who robbed me. That is the guy who *didn't* rob me.' Just like that, when earlier they had told us who it was. Witnesses may come down here four, five or six times. And what do they get out of it? They'll say to me: 'But he'll be out in no time,' or 'He was already out on pretrial release.' So I have a standard little speech I give them. I say we have a very good case. If it's suddenly dropped, he'll know why. He'll know you refused to identify him. That's just like writing 'pigeon' right on your forehead. Anyone want to rob someone, come and rob me."

She lit another cigarette and went on: "The citizens don't understand, either, especially the ones who are poor but honest. I had a case once, a victim in an armed robbery. A blue-collar worker. By the time of trial he had been laid off his job. The defendant was convicted, but he had no prior convictions. Then, pending sentence, he was released by the judge on condition that he take a certain job, which the judge had arranged for him. Of course, the victim would have liked a job too. The defense attorney later told me that the kid quit the job a few days later—it was hard work. But he still got probation. So what does the victim think? He had to come down here a number of times, no job. . . ."

"You know, if a Martian were to come down here he wouldn't believe it," Cross continued. "A crime is committed, right? What happens? In some cases the witnesses are locked up for their own protection, the alleged criminal goes free. There are now such things as 'witness protection units,' used primarily for organized crime cases, but sometimes with street crime. Then the case goes to trial, and you lock up the jury. Then he's convicted, say. At last he goes to prison—maybe.

"But to answer your question about the Sally Ann Morris case, if the defense lawyer can persuade the judge that there is no danger of flight, then there is no reason for not releasing the defendant on his personal recognizance. If they've lived for years in the District, the judges tend to be impressed that they have ties to the community. I've heard lawyers defending multiple offenders tell the judge with a perfectly straight face, 'Your Honor, my client has been charged 35 times, and he has shown up in court every time.' Their long record is used as an argument to release them."

"One more point before you go," she said. "Catch-22 for prosecutors. Say a man is convicted of armed robbery. He's a first offender, and he gets probation. Out on probation, he's arrested again for armed robbery. What happens? We keep him on five-day hold and ask the judge to revoke his probation for the first offense. But often the judge doesn't. He'll say, 'Let's see what happens in this case first.' He's not been found guilty yet, you see. So that's what they mean when they say your first robbery is free.

"Meanwhile, the judge in the new case will release him again, so that he may be on the streets for seven or eight months pending trial, because to set a bond you have to show likelihood of flight, not danger to the community.

"You should talk to John Hume. I'll tell you a story about him. He was trying a first-degree murder case also involving a rape. Bond was $10,000, but the defendant had made bond. In most cases you only have to put up a tenth of the amount, so the figure is meaningless, but it makes the judges look good because it seems as though they are setting high bonds. The day before the trial Hume's wife had to go somewhere, so they had a baby-sitter come over. Her boyfriend came to see her during his lunch hour, and he brought a friend with him. The friend turned out to be the man charged with murder. While he was in the house he looked around and saw that he was in the house of the man who would be trying him the next day—in the house with his one-year-old son. The suspect was convicted a few days later. He was a refrigerator repairman, who had gone into a woman's house and had found her there alone. He raped and murdered her. He later gave a detailed confession."

Before seeing Hume the next day I went to the Bench Warrant section and looked at the "jacket" on cases 48406 and 48407: James A. Weeks and Roy Wade Weeks, the male suspects in the Sally Ann Morris case. Roy Weeks was listed as living on 10th Street N.W., and as having lived there "for three months, with girl friend." James Weeks had been living on A Street S.E., "for one month." Eunice Walker, an alleged accomplice, was listed as living on A Street for "three weeks." How could such brief stays be construed as evidence of "ties to the community"? And who was the unnamed "girl friend"? I recalled Lee Cross's response on this point: "What kills me is when the person listed as 'friend' is the co-defendant in the crime." Police have been searching for Roy and James Weeks, as well as Eunice Walker and Terry Ann Stewart, since July 11, 1975.

I spoke to John Hume on the third floor of the Pension Building. For the past six months he has been head of the Felony Trial section in the U.S. Attorney's office.

"Of the felony cases that we have indicted," he said, "we have 500 fugitives right now. That is 25 percent of the cases. There is a variety of reasons. The defendants don't get proper notice, or they say they never receive it. Many of them come to arraignment, learn about the strength of the case and don't come back. Of those arrested on felony charges, I would guess that 65 to 85 percent are back on the street before trial. When I was in Arraignment Court two years ago, we did not get money bond in as many as 25 percent of the cases. When you consider that we convict 75 percent of those indicted, and that about 75 percent of those indicted are on the streets before trial, you've got a lot of dangerous people on the streets of D.C."

About 3,000 defendants are free on bond on any given day in D.C., about 2,250 of whom are statistically likely to be found guilty.

"With a stick-up in D.C.," Hume went on, "if the person is 18 to 22, he can commit two or three or four felonies and keep going down to Lorton for a wrist-slapping under the Youth Corrections Act—six months to a year for even the most heinous crime. Then the Supreme Court finally ruled that the judge can use his discretion about applying the Youth Act in sentencing. Since then adult sentences have occasionally been imposed on youthful offenders."

Hume continued briefly on the subject of lenient sentencing—a defect in the present system of criminal justice as pervasive as the pretrial release of suspects who may reasonably be presumed to be dangerous.

"But it's the American way," Hume concluded. "Instead of making the hard decision, we lower the standards for all. So my criticism of the system is that it tries to do too much for too many. Most of them just aren't going to be turned into law-abiding citizens, but after six months most of them are right back in the community. What you read about in the *Post* and *The Washington Star* is just the tip of the iceberg. There's a lot of human misery all over this city, and mainly because there are people out on the street who shouldn't be."

How did he feel about the accused murderer being in his house?

"Well, it upset me," Hume said." I realized for the first time how the witness feels when he sees that the person who attacked him is back out on the street before the trial."

Hume introduced me to Peter Chapin, a curly-haired man whose office was across the hall and who told me he was in charge of the Felony Fugitive Program. He said there was "a backlog of approximately 530 felony bench warrants ranging from unauthorized use of a vehicle up to first-degree murder." This, he said, was 25 to 30 percent of all felony indictments.

Chapin lit a pipe and reached for a file. "Okay, some egregious cases," he said, "but bear in mind these are not typical cases. I prepared a memo for Earl Silbert, the U.S. Attorney, in April of 1975. Here is one: *U.S. v. Raymond Boswell.* Boswell had several cases pending. Originally he was arrested for second-degree burglary. Personal recognizance was not recommended by the Bail Agency. He had several prior convictions, including rape, unauthorized use of a vehicle, rape reduced to simple assault, and an escape in 1971. The court ordered a $2000 surety bond, which meant he had to pay $200. Okay, in July 1974, he was rearrested on first-degree burglary. On that date the Bail Agency indicated that he had been booked under a different name for a third first- degree burglary. Since he had no previous record under the fake name, he was released on his personal recognizance in that case. The parole board meanwhile required his arrest for violations of his earlier parole. And the Bail Agency could not verify his address. But the court released the defendant on $1000 surety, which meant he had to pay $100, and he then failed to appear for all three charges."

Chapin read out about ten more such cases, and I then went down to the Bench Warrant section and asked to see the jackets on these cases. I was told I wasn't allowed to see them, according to chief deputy clerk Frederick Beane, because I was a reporter, not a lawyer. (This denial of access to a public record was later repudiated by a spokesman for Chief Judge Harold H. Greene.) In another office I found out that Raymond Boswell had been caught, once again, and his case was available for study in Judge Charles Halleck's chambers, where it would soon be coming up for a hearing.

I approached Judge Halleck's chambers with curiosity. There had been publicity about him in recent weeks. A judicial review commission had held hearings to determine whether he was qualified as a judge. Finally they had agreed that he was "qualified," but not "well qualified," and President Ford had duly reappointed him. He now awaited Senate confirmation.

In *The Washington Post* I had read that "Halleck underwent a radical change in philosophy and lifestyle—but not his colorful courtroom manner—in the early 1970s. He divorced his first wife and married a probation worker, grew a beard and moderately long hair, and became, in the eyes of some, as pro-defendant as he once had been pro-prosecution."

Although there had been a lot of implied criticism of judges in what the prosecutors had told me, they had carefully avoided mentioning names. But it seemed clear that Halleck was someone they had in mind. One person I spoke to would only say that, from the prosecutor's point of view, the black judges tend to be better than the white ones, because they have more "street sense." About 80 percent of the D.C. judges are white, as is Halleck.

I asked the law clerk if I could see the Boswell case. Suspicion, barely concealed hostility, was the response. Why did I want to see it? Who did I work for? Why this particular case? Were all the cases I wanted to see in Halleck's court? She went to ask the judge's permission, returned, and said it was okay.

I looked through the file, but it was hard to unravel the legal jargon. Then Judge Halleck himself swept into his chambers (during a trial recess) and unbuttoned his long black robe.

"It's no use your looking through those cases," he told me. "All you're doing is fueling public sentiment against judges. You should be up there on the Hill talking to congressmen if you want to change the system. The problem is we have too many cases: 160 indictments in my court in eight weeks. If we could cut down on the cases and bring the remaining ones to trial within 30 to 60 days, the crimes by those on pretrial release would be cut way down."

What about the violent offenders, I asked.

"You can't predict lethality," he said. "I'm going to co-author an article with a psychiatrist. If a psychiatrist can't tell, how can I? There is no way a judge can predict what a man will do. We have no psychiatric training—it's all a guess."

Taking off his glasses and shaking them at me he said: "The problem lies here: We have too many cases, and so the backlog is too great. We ought to be trying felonies. But with one-third of our misdemeanor calendar we're talking about prostitution—not dangerous; or selling half a pint out of your back door—not dangerous; nickel bags of marijuana—not dangerous. Let's put a stop to that foolishness. Here I'm doing for society what the Baptist minister is supposed to do.

"And they keep indicting them," Judge Halleck went on, his voice rising to a shout. "They keep flooding the court system. You know what they could do to cut down on the backlog?" He asked me directly and waited. "*Shut* the courthouse doors. . . . " He made a great sweeping motion with his hands. "But the Logan Circle liberals who are fixing up the old houses are upset because of the prostitutes and . . ." Here an aide started to tug at his sleeve, but the judge took no notice.

"We can't talk here," he said. "We should go out and have lunch and talk it over. All right. What's the public upset about? Two things. Crime in the street. And too much tax money going out. So there's a freeze on. No more judges, fewer staff people. We judges are the ones that get dumped on. I'm here till seven at night, trying to get caught up." Again the aide tugged at him, but again he paid no heed.

"Here, let me get you a book," he said, darting off to an inner chamber. He came back with two: Norval Morris and Gordon Hawkins's *The*

Honest Politician's Guide to Crime Control, in which the judge had underlined a proposal to decriminalize a number of "victimless" crimes (drunkenness, gambling, drug abuse, abortion by a qualified medical practitioner, sexual behavior between consenting adults).

Finally, reluctantly, he allowed himself to be dragged back into the courtroom. The second book he showed me, with passages underlined in his hand, was *Alice's Adventures In Wonderland*.

Lost Continence

Somewhere around the National Archives Building on Pennsylvania Avenue I joined the March for Life, now in its eighth year. I was glad to be with them, too. Among the forest of placards carried by the long line of marchers (perhaps 60,000 according to published estimates, 100,000 according to march organizers) I noticed one that Robert Herrick or George Herbert might have written:

> *Some babies die by chance,*
> *But none should die by choice*

The good weather that arrived on schedule for President Reagan's inauguration had stayed on for a day or two, although Reagan himself did not come out of the White House to speak to the crowd of abortion protestors gathered on the Ellipse.

The one part of Nellie Gray's speech the television producers seemed to like was the moment when she said that Reagan had been invited but had not shown up. Mrs. Gray, the march organizer, may have made in mistake in mentioning this because it was predictable that the media would amplify it into one more conflict, carrying the subliminal message to viewers that the anti-abortion crowd persists in being unrealistic, whereas Reagan shows signs of sobering up and confronting reality. . . .

On the other hand, I hesitate to second-guess Nellie Gray's tactics. She seems to know what she is doing. But in her position I would simply have said that Reagan had invited the anti-abortion leaders to see him inside the White House after the march. This was more than Jimmy Carter had ever been able to bring himself to do, despite his much advertised religiosity. Carter, spinelessly in my view, confined himself throughout his presidency to the observation that he was "personally opposed" to abortion—in which case he should have done something to stop it. But he was plainly afraid that doing so would lose him votes, or would annoy his supporters. How does he feel about it now, I wonder? Several million unborn infants were aborted during his presidency, and he lost the election anyway.

Someone should have told Carter that a number of U.S. Presidents before Lincoln were "personally opposed" to slavery. As with abortion, one of the reasons they didn't try to do anything about it was the oft-heard defense of slavery that slaves, after all, weren't "fully human."

It should hardly be necessary to point out that abortion is one of the great issues of our time—perhaps, in truth, dwarfing all others. It is estimated that ten million abortions have been carried out in the United States since the Supreme Court's 1973 decision. If you believe that each one of these abortions was murder, then it follows that abortion is by far the most serious public policy issue in our time. Most people in the right-to-life movement do indeed regard abortion as murder.

It is worth considering the political implications of this, as I had the opportunity to do when the march ended. I made my way inside the Russell Office Building, where hordes of marchers were hammering on the doors and patrolling the corridors in search of their senators, some of whom were evidently in hiding for the day. The guards at the doors, for reasons having to do with "fire hazards," were only admitting 50 marchers at a time. One of the guards told me that far more marchers were coming to lobby this year than ever before.

What is the balance of outrage over abortion? How do the rival sentiments on the issue match up? On the one hand, the anti-abortionist believes that abortion is murder. On the other hand, the pro-abortionist believes . . . what? Since pregnancy is not compulsory, and can be avoided by continence, it follows that the pro-abortionist merely believes that those who are sexually promiscuous should not have to accept responsibility for their behavior.

There is a tremendous moral disparity between these two positions. As a result the impartial legislator is likely to feel a more intense pressure from the anti-abortionists. That is why, it seems to me, liberals have made a bad mistake, in purely political terms (leaving aside the morality of it), in coming to the defense of the pro-abortion position. In so doing they have abandoned the moral high ground they so confidently occupied during the civil rights, Vietnam and Watergate period.

How can you become morally indignant defending dilation and curettage? There's no way. Pro-abortionists don't even believe in their own position, when you come right down to it. They will only say that they are "pro-choice." They don't want to hear that horrible word. Not that their libertarian, pro-choice credentials are at all convincing, especially when examined from the perspective of the unborn.

Because of this great asymmetry of conviction—because the anti-abortionists strongly believe in their position and are zealous in promoting it, whereas the pro-abortionists know that in the end they are only propping up promiscuity—it follows that the anti-abortion people are likely to win this battle in the long run. No matter how outnumbered, the zealous will always prevail over the half-hearted. Liberals know this, and it is the superior zealotry of their foes that they fear most. It will be surprising, then, if the liberals don't sooner or later cut their losses and drop the subject. Otherwise, more and more of them will be dragged down to defeat by the terrible millstone of dead fetuses.

That evening, in another part of town, there was a cocktail party. The invitation read: "The Board of Directors of Planned Parenthood Federation of America cordially invites you to a reception to launch the Federation's 1981 Public Impact program." A crowd of well-dressed party-goers, drinking cocktails and nibbling egg rolls dipped in sweet and hot sauce, were gathered in the Monroe Room of the Washington Hilton Hotel. These were health care professionals, the National Health Council, health planning this that and the other: functionaries of the "health care delivery system," in other words. And that doesn't mean delivering babies, believe me. It means a desk job, a secretary, a file clerk and a decent salary deriving one way or another from the taxpayers. It means going to cocktail parties and talking about "the caring professions." Above all it means no messy patients to disturb one's peace of mind. And here they were, "networking," grumbling about the right-to-lifers, and eating egg rolls.

Planned Parenthood is a 65-year-old organization founded by Margaret Sanger and based on the socialist premise that if only the dreams of planners can be enforced, risk and spontaneity abolished, providence pre-empted, evil eliminated, then Utopia can finally be attained. First, however, rights must be asserted, plans promulgated, the public educated, and the taxpayers made to pay the bill. In 1979, Planned Parenthood's income was $140 million, half of which came from government contracts, grants and reimbursements.

"This conference marks the opening move in our year-long agenda to counter those who view the 1980 election results as their mandate to limit freedom in America," said Faye Wattleton, the president of Planned Parenthood, in a press release available at the door. "The fundamental principles of individual privacy and justice are under the most serious assault since the days of McCarthyism . . . self righteous minority . . . imposing their views of morality . . . fear, ignorance, intimidation." Particularly worrisome to Planned Parenthood is the possibility that their federal funds (made available under the Public Health Services Act) will be cut off.

Someone introduced me to Faye Wattleton, a tall and somewhat glamorous 37-year-old black lady. "Some people simply don't know the facts of reality in the real world," she said when I brought up the voluntary nature of pregnancy. "People do find themselves pregnant. They become involved in a relationship that results in sexual expression. Some people want to stamp out sexuality, which is very much part of us. We can't simply repudiate others because they don't conform to our perception of what their behavior should be."

She turned aside and briefly spoke to a Congressman nearby. I could hear one or two people in the background murmuring against me: What was I doing here? How did I get in? I wondered if they would have the nerve to ask me to leave, but they said nothing directly. Ms. Wattleton turned back to me and said: "It is our desire that no pregnancy be unwanted."

But does that mean that "unwanted" babies are better off dead? (A possible pro-abortion-rally chant suggested itself: "Better off dead than un-want-ed.")

"I have talked to people who wish they had not been born, because they felt they were not wanted," Faye Wattleton responded. I forgot to ask for their names. They would have been worth interviewing—the hitherto unheralded victims of anti-abortion legislation.

"I might point out that the unborn do not have civil liberties under our government," Faye Wattleton callously added. Obviously she enjoyed "sensitivity immunity," safely shielded by her legally privileged "black-woman" status. She forthrightly said things that would give the rest of us pause. "And this should not be brought under the perspective of religion," she added sternly.

I went in search of a drink, and by the bar encountered Joseph F. O'Rourke III, who was dressed like a priest, with black shirt and clergyman's collar. He was holding forth airily to a circle of mostly female admirers. Cigarette in one hand, vodka on the rocks in the other, half-moon spectacles balanced precariously on the tip of his nose, the Reverend O'Rourke seemed please with life and his role in it: media trendy, to be seen on TV quarrelling with the Pope, his anomolous position somehow both demanded and legitimized by the Fairness Doctrine. Today he had been a spokesman for a group called Catholics for a Free Choice—up-to-date clerics who had held a press conference in opposition to the March for Life.

"He was defrocked," said a quiet little voice at my elbow. It belonged to a Capitol Hill aide named Joan Vayo. She was discreetly wearing a March for Life red rose and had come, incognito, to take a look at Planned Parenthood. Obviously she knew the cast of characters better than I.

O'Rourke confirmed the report. "I was a Jesuit priest for 20 years," O'Rourke explained. "I was thrown out for baptizing a child whose mother believed in free choice." Flippancy seemed to be his forte. "Apparently this abortion thing is serious for the Church," he said. "There's no earthly reason—perhaps there's a heavenly one but I'm not in touch with that evidence—why the church shouldn't change its position on abortion. The tradition has been criticized from top to bottom. In every other area of social policy the Catholic bishops have changed."

There was some talk of polls, with the implication that what was formerly regarded as true ceases to be so when a majority expresses a contrary view. O'Rourke told me that he still says Mass. He hasn't been excommunicated. "The bishops don't excommunicate any more," he said, implying that they dare not. "There are many things for which it used to be automatic—speaking out in favor of abortion for example—but now they never do. They always back down."

"That would make martyrs out of them," Joan Mayo told me in another quiet aside. "Look at what happened to Sonia Johnson when the Mormons excommunicated her. She became a martyr."

But Sonia Johnson's celebrity was short-lived. In a recent article on Ms. Johnson, *The Washington Post* attempted to inflate the conflict between her and the Mormon Church by saying that she was only "technically" not a member. The paper thereby paid inadvertent tribute to the role that membership plays in legitimizing changes in any organization's teaching or purpose. Outsiders can hardly expect to alter a group's by-laws. But the views of "dissident" members do enjoy at least some legitimacy. When John Paul II visited America in 1979, NBC gave O'Rourke 30 seconds to attack the Pope without even pointing out that "Father" O'Rourke was no longer a priest in good standing. But he *could* claim still to be a Catholic. Unmartyred he may be. Unexcommunicated he is free to muddy doctrinal waters.

O'Rourke called for another vodka. By now most guests had left. Suddenly he flipped the clerical collar out of his shirt. It turned out to be nothing more than his calling card, neatly folded lengthwise and inserted between the wings of his shirt collars. He undid the top button of his shirt and lit a cigarette, transforming himself into a civilian. The day's work was done.

"Hi, gorgeous," he said to a Planned Parenthood female functionary, who had come over for a chat.

SONIA JOHNSON'S CONSTITUTIONAL

One hot day in July, 1982, shortly after the defeat of the Equal Rights Amendment to the Constitution, a group of strange-looking women held a demonstration in front of the National Archives building in Washington, D.C. This is where they keep the original copies of the U.S. Constitution, the Declaration of Independence, the Bill of Rights, and many other treasured documents. I happened to be walking nearby, down Constitution Avenue as it happened, just as the women began drawing blood out of each others' arms with hypodermic needles and mixing the blood with red paint.

There were 30 or 40 of these women, mostly with the wizened, sun-baked look that comes from a steady round of outdoor marching and demonstrating. Many of them were wearing scarlet T-shirts. They started squirting the blood and paint from plastic ketchup containers onto the building. The great Greek columns of the National Archives began to take on the appearance of a Jackson Pollock painting in its early stages. A stone statue nearby, representing the Heritage of the Past, was liberally doused by a woman who was herself paint-streaked. She was standing in the statue's lap and squirting away with her ketchup container like a wilful child defying her parents to spank her.

I wondered if I should make a citizen's arrest. On becoming an American citizen I had raised my right hand and sworn to uphold the Constitution. Here was a challenge to it. Finally I spotted a policeman. "Is this legal?" I asked.

"I've got my orders," he mumbled disconsolately. He looked as unhappy as a guard dog who knows that someone unauthorized is at the door but has been carefully trained not to bark.

"This land has been stained—by the oppression of women."

The women were now standing in a phalanx on the Archives' steps, swaying from side to side, arms up in V-formation, and they were chanting in unison, over and over again, "This land has been stained—by the oppression of women."

A burly woman with an armband picked up a megaphone and announced: "The rally will continue at the Department of Injustice." She was referring to the building just down the street, once the home of J. Edgar Hoover before the FBI moved into its own building on Pennsylvania Avenue. By now there were tourists and media; Hitachi cameras were going clickety click. The protestors moved on to their new target, now chanting: "Wimmin fight back, wimmin fight back." Most seemed to be in their 30s or 40s. They looked as though they might very well have been veterans of the Vietnam War (domestic version).

"That's a real mess," said a passing tourist, looking up at the Jackson Pollock Archives.

"Lady," replied a demonstrator, "the whole archives are a mess." She was referring to the male-chauvinist documents, the Founding Fathers, the sexist Constitution, and so forth. The demonstrators had made a "statement": The outside of the building should resemble its contents.

"Oh, look at that," said another bystander. "I was for the ERA until you sprayed the Archives."

After the women had moved on down the street, a different policeman appeared. He was dragging a shiny black garbage bag down the steps, occasionally stooping to recover stray leaflets and Coke cans. He did a nice job of cleaning up for the ladies.

Now a dozen of the women were holding hands in a line right across Constitution Avenue, blocking off traffic. At this point the reluctant police seemed to realize that they would have to take action. A cop with a bullhorn read the law to the lawbreakers, advised them of their rights, went through the entire Miranda procedure on Constitution Avenue. After elaborate patting down for weapons and much filling out of forms, the traffic blockers were led off one by one into a paddy wagon.

You could see the women sitting together in the back of the van, holding hands, swaying from side to side and chanting: "We will never give up, we will never give in." They were enjoying the briefly experienced trance of collectivism, their individuality submerged, their identity lost, their troubles forgotten for an hour. They might as well have been hypnotized.

"Get all the way up on the curb there . . . " a policeman was warning someone. To which a woman near me responded: "Well, there goes the First Amendment."

I asked several women what their names were. But they refused to oblige. They were friendly, though, regarding me with my notebook to be "media," therefore on their side. One demonstrator patiently instructed me that they were to be referred to as "a group of women, lower case and no quotes." Another was wearing on her T-shirt a button reading: "Nuke A Gay Whale For Christ."

"What does that mean?" I asked.

"It's the all-purpose right-wing button," she confided.

Twelve women were driven away in the paddy wagon. The remnant gathered in a kind of witches' circle on the grass beside the Justice Department. One of their number delivered the group's formal statement: "June 30, 1982, will live in infamy," she read. "In one of the most barbaric decisions of our history, a handful of male legislators in unratified states misused our democratic process to keep the Equal Rights Amendment from ratification. Those men chose greed and profit reaped from the labor of poor women over justice and human rights."

I spotted Sonia Johnson among the women and I managed to speak to her briefly. The Mormons had excommunicated her for agitating for the ERA within the church. She then wrote a book with the wonderful title, *From Housewife to Heretic*, setting forth her "growth"—her disavowal of traditional social constraints. Note that "heretic" has become a term of approval within the present liberal culture; a compliment, a label to be embraced.

Sonia Johnson was in Illinois for the final ratification struggle, publicly fasting in the state capitol for 37 days along with several other women. All were dressed in the noble white of martyrdom. The news media duly fussed on their behalf. Sonia lost 22 pounds. Now here she was drinking a Coke on Constitution Avenue, thoroughly enjoying the renewed attention.

"Did you really say you'd like to strangle God?" I asked.

"I said I'd like to kill him," she corrected me, sipping her soda. Camera crews from local television station were jostling up close behind me, microphone booms swinging in over my head.

"Would you take your sunglasses off," someone from the media asked. She was happy to oblige.

No, she went on, she didn't believe in heaven, and she didn't believe in hell. The next life, if there was one, was something that would take care of itself when the time came. *Now* was the hour, the hour of great struggle against men and against all traditional beliefs. In fact, the "movement" of which she was a part had only just started. "I am very happy to be alive at the beginning of it," she said.

I went back down the street, and there was another movement of women on the Archives steps—charwomen on the move with scrubbers and bristly brushes and Ajax cleanser. Scrub scrub scrub went the charpersons, while down the street Sonia Johnson and her friends continued their protest against injustice for the benefit of the cameras.

THE DALY WOMAN

Last month I went down the road to hear a campaign speech by a presidential candidate—our old friend Sonia Johnson, who has been selected to lead the Citizen's Party. (In 1980 their candidate, Barry Commoner, was on the ballot in 30 states and received 236,148 votes.) Sonia was speaking at the University of the District of Columbia. I had not realized that such a place existed, although it is little more than a mile from where I live. It attested, in my opinion, to a local superfluity of tax revenues and ever more implausible excuses to spend them: a cluster of cumbersome concrete structures, outside which stood vertically inscribed signs (so that you have to twist your head to one side to read them), usually containing the word "resources" in the title.

Somewhere underground I found a luxurious auditorium. In the hall outside tables were set up, behind which sat women selling ideological books and pamphlets—for example, Marianne Wax's *Let's Take Back Our Space: 'Female' and 'Male' Body Language as a Result of Patriarchal Structures* (with 2037 photographs).

There must have 300 or 400 of us gathered in the auditorium—perhaps ten men, and almost no blacks of either sex. Was my "space" a tiny bit threatened? I couldn't help wondering how many of the wimmin who came striding down the aisles in bib-overalls were enrolled in the karate course advertised outside.

The warm-up speaker was called Mary Daly, and she would be reading from her new book, *Pure Lust: Elemental Feminist Philosophy*. In fact, the poster advertising the event had Daly's name in lettering larger than Sonia Johnson's. On the cover of her earlier book, *Gyn/Ecology*, Ms. Daly is described as an "associate professor of theology at Boston College," and a "Revolting Hag." I gather that she was once a Catholic and perhaps still considers herself to be one.

According to *Ms.* magazine, Sonia Johnson became "instantly internationally famous" when she was excommunicated from the Mormon Church for supporting the Equal Rights Amendment. In 1981, Doubleday published her book *From Housewife to Heretic*. As *Ms.* put it, "No one could have seemed less likely than Sonia Johnson to become a famous activist, much less a heretic." In 1983, Sonia ran for chairperson of the National Organization of Women but narrowly lost.

Now she came to the podium—a slight figure wearing a cerise jacket over a mauve top, plain black trousers and a lapel button reading: "Listen to a Woman for a *Change*." She told us how "moved" she had been when she found that Mary Daly "recognised and approved of what I was doing." She seemed to regard this Mary Daly as being altogether her superior. It was unusual to find the main speaker introducing the warm-up speaker. Daly's poetry, Sonia said, not only exposed the "motives and machinations of the masters," but helped us to "remember our deep ontological wilderness." And so she was honored to present the great Daly Woman to us.

She had short grizzly hair and she wore a horizontally striped rugger-bugger jersey. If she came walking toward you on the sidewalk, you might just decide to step aside rather than risk invading her space. If women played football, she would make a good linebacker, or whatever it is they call them. Right off she struck the microphone with the palm of her hand—Bonk!—and said: "These little phallic things! Sometimes they work and sometimes they don't!"

The audience cackled its appreciation. The Daly Woman then gazed out at her admirers and asked: "Are there any Revolting Hags or spinsters here?" I suppose this might have been a request for lesbians to identify themselves. There was a sizeable show of hands.

". . . All of our fore-sisters, past and future, so we are gathered in 1984, a period of extreme danger for wimmin, extinction by nuclear holocaust or chemical contamination which wimmin are living through to sustain a biophilic consciousness . . . dragon-identified. Pure lust does journey through three realms, the realm of archospheres, pyrospheres, recovering the volcanic virtues of *wild* wimmin. It's 1984, the time of Reagan & Co. So, looking at the sado-society, penocracy, the product of the phallic flight from lust, transforming it into the more sublimed phallic lechery—that's what we call civilization?"

"Hooo hooo hooo hooo hooo!"

Oh, the monstrous regiment that was congregated loved every minute of it, every pun, every jeer, every quip. Ms. Daly proceeded to ridicule St. Jerome and St. Simeon Stylites, the former having been visited by "a bevy of dancing girls in his imagination," the latter attended by a flock of faithful who "gathered the precious turds that dropped from his body." She ridiculed the Pope, and she ridiculed St. Paul, from one of whose Epistles she read in a tone of sing-song sarcasm, ending triumphantly: "Well, speak for *yourself*!"

"Rage is like the volcanic dragon-fire," the liberated theologian continued. Hard to believe that she actually teaches at an American university! And teaches theology. "This rage is what moves us into the race of wimmin . . . The whole world is groaning under phallic rule." What is needed, she explained "is the courage to sin," which is simply "the courage to be," according to her etymology.

She mimicked her critics, speaking in hilarious, mincing tones of ridicule. She took particularly perverse pleasure in impersonating the middle class housewife who might have strayed by accident into a Daly discourse. "'Oh, she's so bitter,' they say. 'So bitter.' These are the people who talk about 'fulfillment.'" She repeated the word prissily on the tip of her tongue, fastidiously enunciating each syllable, thereby drawing attention to the supposed asininity of the word and those who unthinkingly use it: "Fulfillment. . . . Ful-fill-ment. Do you want to be a ful-filled woman?"

"Heh heh heh heh, haw haw haw haw," went the Revolting Hags.

She talked quite a bit about witches, unintelligibly to me—although I do now understand why they burned them. She made a blasphemous joke about "God the flasher, God the stud, and God the holy hoax," and then introduced Sonia Johnson, who was plainly thrilled and awed by Daly's boldness. Daly told us not to worry about voting for Sonia on the grounds that this would "take a vote away from Mondale—as though we should be grateful for slightly lesser evil! They'll punish you for being a little itty-bitty feminist. So why not go the whole way?"

Sonia began by telling us that she could no longer remember how to spell her last name. Everyone understood perfectly what she meant. The Woman's Movement, she proclaimed, was the "greatest movement in the history of the world," and it had come "in the nick of time." You see, she had been caught red-handed in her own sexism, earlier on, and so why was she laughing? I mean, look at who *is* President. Someone on the plane had suggested that she run for President—she was going to see Barbara in Florida who had been so supportive. . . . So-o-o, the minute she started to think about it, she *knew what to do*, as President that is. She realized that they had mystified the office, deliberately of course, but it was really clear now and she knew what she'd do on her first day in the Oval Office. Think about it, audience, do *you* know what Sonia would do? Let's hear some of your suggestions for what you would do.

Sonia lifted up her arms like a conductor facing a familiar orchestra, so that it was only necessary for her to make the slightest move of her baton to elicit the expected, indeed inevitable sounds.

"Disarm all the missiles!" someone shouted out.

"Write a new dictionary!"

Sonia bounced her invisible baton, or wand, to encourage the diffident wind and strings.

"Clean up the environment!"

"Institute 24-hour day care!"

"Pass the ERA!"

"End all dealings with racist South Africa!"

Sonia was well pleased with the response of her players. "I didn't hear anything that isn't in my platform," she said. "Feminism is the most all-encompassing analysis of the human situation that we know of," she noted—acknowledging the power of an ideology to organize the opinion of its adherents. "If we had a feminist in the White House, the chaos would disappear."

Margaret Thatcher and Jeane Kirkpatrick were mere "female impersonators."

We should: demilitarize the planet, cut the military budget, unilaterally disarm; cooperate instead of compete; stop brutalizing the planet, and stop raping and poisoning our embryonic fluid; use winds and sun and biomass; and "work with Earth's rhythyms."

Sonia said that with a little more effort she would be getting Federal matching funds. Don't-you-love-that? "I love the fact that the Federal government is going to pay me to say it," she said. And what's more, she believed that she might be included in the League of Women Voters' debates.

Someone had asked a question about racism. Here was how Sonia would address the problem. "One of the first things to undermine the racist mind, the President can make 3000 appointments—the Secretary of Defense and the Federal Reserve Board, places with *enormous* status and where money is taken care of. And you can begin to undercut homophobia by appointing gays and lesbians—you *immediately* attack the problem at the root. Not just token places. Powerful places like the Supreme Court!"

Barbara Demming had said: "We should run a coven for President. Convene covens all over the country! It would be a lot more democratic. A lot less hierarchical. Not just revolution," Sonia explained. "Transformation."

Someone asked a question about Congress.

"Congress, what are we going to do about Congress?" Sonia agreed. "Just when we need them, they go flaccid on us. . . . "

Oh no! There was a low groan and the audience looked crestfallen when that metaphor sank in.

"Oh, that was an awful joke," poor Sonia agreed. She would do better and she would try again. With a different metaphor, let's hope!

"They . . . they don't rise to the occasion."

Oh groan. No no no. Take it back.

"They just . . . they just keep doing it!"

Well, how very Freudian my dear. Fatally trapped, you see, by her patriarchal, sado-social, penocratic, life-hating upbringing. It was understandable. It demonstrated the very point they were all trying to make. Ideology can show you the way out, but conditioning can keep you in the trap if you're not careful. Nonetheless, Mary Daly would have avoided these gaffes. I could see why Sonia deferred to her.

"This is getting really funny," Sonia worried. She made a desperate escape from her male-conditioned trap by saying: "We're just going to have to restructure stuff a lot. Have to remember that this country was founded by a patriarchy."

By the exit they were selling a "Wanted For War Crimes" poster of Reagan, a poster prepared by the Citizen's Party. The four listed crimes were: Invasion of Grenada. Overthrowing the Nicaraguan Government. Killing of Civilians in El Salvador. Deploying First-Strike Nuclear Weapons.

A blurred, unformulated question nagged at me as I made my escape, threading my way between the Revolting Hags and the concrete Resource Centers that comprised the university. Oh yes . . . that was it. Why was it, again, that they had found the Democratic Party to be inhospitable to their views? Sonia Johnson had said why, and it seemed reasonable at the time, but now that I was out in the open air I could no longer recall her argument.

The Culture of Washington

LUNCH WITH THE ROCKEFELLERS

For months I found it hard to understand why Washington is somehow so grey to the mind's eye. The weather tends to be sunny, the sky blue, the atmosphere clear, the surrounding countryside as green as you could wish.

For unrelated reasons I called an old Washingtonian, Julia Cameron. I had met her in New Orleans a year or so earlier. She had moved to Hollywood and married the movie director Martin Scorsese, had a child by him, then suddenly divorced him. I asked her how she liked Hollywood.

"There are so many pastel shades," she said—out of the blue.

Of course. One thinks of Hollywood in technicolor.

"Why is Washington so grey?" I asked.

She thought for a minute and said: "It's the newsprint."

Of course. What a beautiful perception. More or less everyone in Washington has a professional obligation to read the newspapers. They are our trade journals.

A few days later, in the final days of the Ford Administration, I went to the Sans Souci Restaurant, having heard that this was an authentic Washington experience. It is a kind of theater in the round where one may see and be seen by the famous people of the moment. Simultaneously one eats lunch. By remarkable good fortune my friend Virginia ("Vivi") Harrison and I were seated right next to Vice President Nelson Rockefeller, who was having lunch with his wife Happy, and a local hostess, Joan Braden. Rockefeller would be Vice President for four more days.

The humorist Art Buchwald was also in the restaurant (he was to be seen there every day in that period) and it wasn't long before he came brushing past us to exchange gossip and jokes with the Vice Presidential party.

After he returned to his seat Joan Braden and Happy Rockefeller began exhorting the Vice President "not to be afraid to speak out—really tell the people what's on your mind." As a former Vice President he would have a media platform, they implied, and the people would be sure to listen. So—he would be able to get his message across. But I couldn't figure out what this urgent message was, exactly, and I suspect the Vice President may have had his doubts, too. Either that or he had fewer illusions about the people. I watched Rockefeller as they encouraged him to Go Public. He was leaning back in his chair, occasionally forking spinach into his mouth, with the back of his left hand held under his chin to prevent spillage.

"Yeah, yeah," he said, not really interested. He looked for all the world like an elderly gentleman lunching at the Yale Club, or wherever it is that establishmentarians have lunch in New York. There was some resemblance, in fact, to pictures one had seen of his remarkable grandfather, John D., the one who made all the money.

By his demeanor he seemed to be saying . . . not that he was *afraid* to speak out exactly . . . but that somehow he knew it to be a quixotic venture. To be a Rockefeller in the last quarter of the 20th century! A symbol of the capitalist class after the arrival of the New Class; like a surviving dinosaur after the mammals were already well ensconced.

Lunch dragged on. By three p.m. we were all still in place, with the exception of Art Buchwald. The Rockefeller party seemed to be making an afternoon of it. No great pressing business for the Vice President, evidently. Unexpectedly, Happy Rockefeller turned to V.V. Harrison— there was barely two feet between them—and included us in the conversation. Who were we? Happy wanted to know, and what did we do? She said something about her children. A son might turn out to be an athlete: a Rockefeller of the track. And why not? For him the board room might seem somehow . . . not quite right.

"My husband quotes Darwin to me all the time," Happy said at one point. She turned to Nelson, who was drowsing peacefully after his meal. "What is that about Darwin you tell me?" She said.

He perked up. "To survive, you gotta adapt," he said.

At that point I began to suspect that his lack of enthusiasm for "speaking out," as his ladies-in-waiting had advocated, was borne more of confusion than futility. Was he not the dinosaur who had aspired to be a mammal? As a result he didn't belong in either class and he knew it. If he were to speak out, what would he say? One sympathized with the old fellow.

When their check arrived, Joan Braden made a manful play for it. But the Vice President reached for it just in time, with his other hand deftly slipping a gilt-edged wallet out of his breast pocket. In the struggle for the check, if not the struggle for existence, Nelson Rockefeller showed himself to be well adapted.

Jimmy Carter has already performed one signal service. He has effectively brought the Kennedy dynasty to an end. The other day I met the prominent Washington journalist Sally Quinn, who has made herself both famous and unpopular in Washington by dissecting the techniques of social climbing in the nation's capital. This she did spectacularly in a long article about a social newcomer, almost completely unheard of, called Steve Martindale, in the "Style" section of *The Washington Post*. The article included a list of do's and don'ts for social climbers. Quinn is an interesting figure, a dry and sardonic observer of social and status detail. She is married to Benjamin Bradlee, the *Washington Post* editor of Watergate fame and formerly a close friend of President Kennedy.

Quinn, when I met her, was discussing the great resentment on the part of the remaining Kennedys, notably Ethel, toward the *arriviste* Carters from Georgia: "Jimm-ah and Rose-alynn," and at this all present indulged in great guffaws—both at the impertinence of the rustic newcomers from Plains; and, somehow (because it was too delicious) at the shocked, how-dare-you reaction of the fashionable residue from Cape Cod (who thought the White House was theirs by hereditary right). "The Kennedy people all voted for Ford you know," Quinn said. How marvelous! And how hollow the ideals of Camelot, that they could be laid aside for four more years, in the hope of a Restoration in 1980!

One of the things that makes Washington interesting is precisely that it is so "open" socially to newcomers. There is no suspicious, fusty old aristocracy, as in New Orleans, which doesn't recognize the existence of those whose parents weren't born in the city. Newcomers are actually welcomed in Washington as an amusing diversion. The quadrennial elections ensure an endless supply of new talent.

THE ELECTRIC WINDMILL

Curious about the overnight appearance of wigwams and other quaint structures, I parked my car near the Mall. A whirling windmill was also to be seen on the grassy sward, not too far from the Lincoln Memorial. Were the Indians in town, putting on one of their periodic Media Events? Worth a look at least. On closer inspection it seemed to be a fair of some kind. Semi-naked youths were strolling about and lounging in the grass. Not the Indians, but the hippies back with us, it seemed.

Someone handed me a press release. I had stumbled upon the Appropriate Community Technology Fair, called ACT '79, a "celebration of old-fashioned American ingenuity." Reading further, I learned that the fair, "a self reliant, environmentally clean and democratically governed instant community, will simulate the sights, sounds and other sensations of a real community." This was the Small-Is-Beautiful Crowd. Slanting solar collectors were dotted about. I kept a wary eye open for Amory Lovins or Barry Commoner, and was ready to dash for cover if either should appear on the scene. The spirit of E. F. Schumacher hovered uneasily over the sward.

I followed a footpath between tents, inside which seminars were in progress. I stood in the back of one and listened for a few minutes. All of the instructors at the other end of the tent worked for one or another government agency. They were sitting in a row behind a table and talking happily away about viable options, one or two of them intermittently taking meditative little puffs on their pipes. Coordinators, moderators, biodegradable resources, renewable coalitions, recycled neighborhoods, community-run revitalization projects. Puff, puff, puff. It beat staying indoors all day, imprisoned in the Federal Triangle.

I blundered into the WomanSpace tent. Importance of coalition-building, resource recovery noted; poverty and the Third World Woman; post-patriarchal responses to the world predicament considered.

I strolled out into the sunshine, which was energizing a solar collector, which provided power for a record player. Lovely. Hippies and layabouts lolled on grass listening to rock music, at last independent of the ripoff oil companies. But if perchance you stand in front of the solar collector, the music runs d-o-w-n h-i—l—l, and then grunts to a halt, and they look up at you and moan, "C'mon, man. Give us a break."

Did these people ever do a day's work in their lives? Ten years ago daddy's credit card paid for everything. Now they imagine the sun is going to put in the effort on their behalf. They preach self-reliance, but they practice little.

I made my way past herb growers, people sitting like children in magic circles, tepee dwellers, woman's community bakery, bio-gas model project, "Why-Flush" water quality. Heard the word "retrofit" used a dozen times. Eventually reached the Administrative Tent. Asked for "the boss" and several people immediately looked up and turned to stare at this relic of the old order.

No bosses here, I was told. In the wrong place, man. This was democratic. No bosses in Lovinsland. But someone called Bob Zdenck appeared eventually and told me how the fair was put together.

"First we wrote a proposal for funding from the National Science Foundation, the National Center for Appropriate Technology, Housing and Urban Development, Department of Labor, Department of Energy, Department of Commerce, and the Community Services Administration," he said. "In September, 1978, we got a $19,000 planning grant from the Department of Energy. We hired Michael Duberstein in October to begin planning and logistics. Then we hired two outreach workers and an administrative assistant.

"The next thing was we got a $17,000 grant from ACTION [a new-ish Federal agency that includes the Peace Corps]. We held a very important meeting with Bill Holmberg from the Office of Consumer Affairs, Department of Energy, and he agreed to help us with a lot of in-house cost. Then we got numerous other grants: $15,000 from the Economic Development Administration, $15,000 from the National Endowment for the Humanities, $15,000 from the National Center for Appropriate Technology, $5,000 from the Small Business Administration, $2,500 from HUD, and a lot of in-house from the National Park Service. We have a core staff of five people. But the Department of Energy is paying for others. The Baltimore County CETA Consortium made the building facades. And we had public service announcements on television in ten states."

I asked Bob where Michael Duberstein was to be found. "Showing Senator Tsongas round the fair," he said. Tsongas! So! He too was implicated. I thought he might have taken up residence in Tanzania by now, the better to act as Julius Nyerere's public relations man. Didn't know he was involved in this solar malarkey. Next thing you knew, Ralph Nader himself would come loping down a path with a file folder under his arm, wearing his conscientious frown.

I decided to take a look at the windmill, a large three-bladed propeller on top of a tall tower. The propeller was churning around merrily, although there was little or no wind at ground level. On the way I stopped at the "bio-gas" demonstration and was informed that the people here received a Community Services Administration grant of $150,000. (It was beginning to look as though Bob Zdenck had underestimated the taxpayers' unwitting contribution.) At the foot of the windmill a rather well-bred Vermonter was explaining everything about the contraption. The windmill cost about $4000 to buy and install, he said. It would save about half your electricity bill—*if* you lived in a windy spot. Using his figures, I concluded it would take at least 20 years to pay off the investment—*if* it never broke down. The prospect of shinnying up the mast to repair worn-out bearings was not at all enticing. . . .

Plainly I was contemplating a rich man's toy. Federal tax credits make it less so, however, at the same time encouraging the diversion of capital into economic channels of dubious merit. The Vermonter waxed enthusiastic about Wisconsin's progressive congressman, Henry Reuss, who had installed a windmill in his backyard. I believe he also pushed through the tax credit.

I asked the gentleman from Vermont why the blades were whirring around so smoothly in such still air. "It's not working off the wind," he said. "It's plugged into the power outlet." It wasn't demonstrating the production of electricity. Electricity was demonstrating it.

Somehow, at that moment, the sun went in. And the rock music stopped. But the windmill went on turning, and the Federal money, I am sure, still pours down on these artful dodgers of the 1970s, the residue of the counterculture, who have discovered that Big Daddy in Washington has the credit card now, and is ready to put it at their disposal until they decide what they want to do when they grow up.

KENNEDY'S HEARING AIDES

Wherever Senator Edward M. Kennedy goes, he is surrounded by a tight cluster of petitioners and protectors who maneuver for positions around his bustling person. "Senator, senator . . ." some call out, as they trot along in his wake—the whole convoy proceeding at well above normal walking speed. Others carry briefcases and wave memoranda. Still others—Secret Service, for Kennedy is campaigning for the Democratic nomination—have little insignia in their lapels and wear large cut suits to accomodate the guns under their armpits.

Here they come, Kennedy leading by a nose, rounding corners and now clattering down the final straight, which is an echoing flagstone corridor in the center of the United States Capitol building. Kennedy is on his way to the Senate Judiciary Committee hearing room, where he will preside over a "markup" session of Senate bill 1722.

A markup session is what happens when the senators on a committee get together and debate among themselves various proposals to amend pending legislation. S. 1722 is an elaborate attempt to rewrite a section of the criminal code. It has been winding its way through the legislative labyrinths for eight years at least. The draft bill is over 400 pages long, and very few people in America have any idea what the whole thing means or why it is necessary to rewrite the criminal code (if indeed it is). Judging by their reactions at today's markup, quite a few senators don't either.

The hearing room is plainly too small for the number of people in it. For some reason the usual Judiciary Committee room is closed today. This is more like holding hearings in a wine cellar. A long table with microphones in front of each chair awaits the senatorial arrivals. The room is already packed with senators' aides and reporters. The reporters carry notepads, but are probably more interested in Kennedy than in the bill under consideration. The numerous legislative assistants, who outnumber the senators two or three to one, are more interested in trying to influence the bill this way or that. It is true to say of most of them, if not all, that they spend a good portion of their daily lives imagining that they actually are senators, in which capacity they find that they easily outperform their employers.

Kennedy enters the room and pops a cigar into his mouth. Heads turn and there is a subtle change in the noise level. When a celebrity shows up at a crowded party the decibel level rises. But this is more complex: the murmur level rises. A Secret Service man takes up station by the door, and Kennedy moves toward his chair, taking his jacket off. The jutting line of a back brace, like a corset, is plainly visible beneath his shirt.

His face is salmon pink, his hair somewhat tousled. Kennedy has about him an air of slightly disreputable, roguish charm. You get the feeling that were it not for the accident of birth he would today be far, far away from the tedium of a Judiciary Committee meeting—perhaps tending bar in South Boston or rollicking about with half a dozen pals at the race track. It is possible even to feel sorry for him—driven by circumstances at least in part beyond his control into a public life for which he is (I guess) temperamentally unsuited, and which he may not enjoy.

Sitting next to him is the aged Senator Strom Thurmond of South Carolina, an extinct volcano most unlikely to erupt. Very close to Kennedy's right and slightly behind him is a slim, balding gentleman named Ken Feinberg, the special counsel to the Judiciary Committee. His role proves to be that of interpreter. Senator Thad Cochran, an eager-to-please Republican from Mississippi, says something to Kennedy about a proposed amendment, and as he speaks Kennedy has his head turned to the right, listening not to Cochran but to Feinberg, who whispers continually Kennedy's ear. As Kennedy listens he peels off pieces of the outer leaf of his cigar, lost in thought.

Then a gentleman begins to speak in a loud, clear voice from the audience, several paces away from the senatorial table. This is Mr. Ron Gainer from the Department of Justice. There is a respectful hush from the legislative branch as this visitor from the executive explains what the law is and how it is about to be changed. One had thought that the separation of powers doctrine would have prevented such an exchange. Kennedy looks down at his cigar and peels off a few more flakes. He defers to Gainer throughout the proceedings. At another moment he openly seeks the advice and consent of the man from the American Civil Liberties Union.

The eager young assistants are beginning to creep up closer to the senators' chairs, many of which are still empty. After a while the urge becomes irresistible and they openly sit in the chairs. But they know their place, taking care to sit only on the *edge* of the chairs. The senators put their elbows on the table but the aides do not—as a rule. Sometimes a bold assistant might push protocol to its limits and put *one* elbow on the table. Kennedy in any event is not bothered about such details. He welcomes any and all comments. There appears not to be an authoritarian bone in his body. He resembles nothing so much as the easygoing headmaster who lets the students rag the faculty.

"Ah, Mr. Schwartz, what's your reaction on this?" he inquires at one point. Mr. Schwartz, wearing heavy horn-rimmed spectacles, is sitting with an air of great dignity and comfort in the chair of Senator Metzenbaum of Ohio. It is not clear whether Mr. Schwartz is from Ohio, but he gives his reaction.

Now Senator Birch Bayh of Indiana makes his entrance—chewing gum. He stands by the door for a while, looking out over the assembled company as though he were, say, Marlon Brando, continuing his chewing, as if to say: I'm just going to stand here and go right on chewing gum for a while and if there's anyone who doesn't like it, he can come on up and tell me and we'll settle it outside the room. If Kennedy is the gregarious headmaster, Bayh is the football coach who knows a thing or two about discipline.

Finally Senator Bayh takes his chair, and two aides behind it get down into the approved "crouch" position, ready to start *their* whispering (for Bayh too will need interpreters). But within a few minutes the Indiana senator gets up again and leaves the room. He is restless today. Then the roaming coach returns, this time holding his coat over his shoulder, Camelot-style, and again he stands in a conspicuous spot and surveys the company.

There is some obscure talk, started by Senator Simpson of Wyoming, about the First Amendment rights of protestors occupying nuclear power plants. Lean of visage, Simpson enjoys the country-lawyer role popularized by Senator Sam Ervin at the time of Watergate. Simpson is someone other senators look up to—he seems worldly wise and yet in touch with the people: a combination to be emulated.

"I wish the Justice Department might be able to help," says Kennedy, who is playing with his gavel. "Can you comment on that kind of concern, er, Ron?"

He can, and for good measure Bayh adds that "the First Amendment doesn't give anyone the right to go into my backyard or into nuclear power plants."

But no one in the room is against the First Amendment either. Rest assured of that. "We're not trying to run a whizzer so that we crush First Amendment rights," someone reassures.

". . . grey areas . . ." Kennedy concludes.

Bayh has a peculiar amendment he wishes to insert. It would permit wives to charge their husbands with rape. He is opposed in this by Senator Orrin Hatch of Utah, who has been sitting through the proceedings with an air of long-suffering rectitude. Hatch worries that the Bayh amendment will undermine the family.

In his back-and-forth with Hatch, Bayh is advised by Jessica Josephson, who is tall and wears a mid-calf tweedy skirt and matching jacket. In whispering to Bayh, she won't assume the crouch position, perhaps thinking it too undignified. Instead she executes a quick little knee-bend to lower herself to Bayh's ear. From across the table it looks like a curtsy. While Hatch speaks she gives him an aggrieved look, as though his comments were directed at her personally. In response she gives another whisper-curtsy, and Bayh responds.

Hatch eventually gives up the battle when Senator Thurmond makes it clear that he is not offering any support. "The Senator from Utah is right about 99 percent of the time," Thurmond suspects, but this is not one of them. So Hatch withdraws his objection. Miss Josephson gives one last curtsy, and the football coach enjoys his moment of glory.

All good things must come to an end, and before much more progress is made Kennedy picks up his gavel and announces that the committee will reconvene for further deliberations on the criminal code the following Tuesday morning. Kennedy puts on his jacket, and an appropriate amount of hand-grasping, nodding and smiling, murmuring and plotting is carried on around him. Finally his entourage gathers itself together and sweeps him out of the room, out into the slipstream of the presidential campaign.

LETTER FROM MARTHA'S VINEYARD

On the third day I was stung by a jellyfish named Bitsie or Muffie, trailing an "Anderson for President" sticker from a tentacle.

The year-round population of this 100-square-mile island off Cape Cod is only 9,000 or so—a good deal smaller than it was in its glory days of whaling in the first half of the 19th century. But today the summer population swells to about 70,000, mainly because, in the words of Art Buchwald, Martha's Vineyard "is the 'in' island at the present."

The humorist continues: "Those of us who have been going there over the years never planned it that way. As a matter of fact, many people fell in love with the island because it wasn't 'in.' We sailed, fished, clammed, and tended our lobster pots, oblivious to the goings-on in such Gucci-ridden places as Southampton, East Hampton, Newport, Bar Harbor and Malibu."

Local residents question whether Buchwald ever tended a lobster pot, or so much as clammed on the island. As for being oblivious to the goings on in East Hampton . . . I mean, you just know he hasn't figured out the socio-dynamics, if that's what he thinks. The point is that the fashionable folk are quite happy to abandon the Gucci-ridden Hamptons *provided* there's a whole rat pack of them ready to migrate at approximately the same time—and to the same rendezvous. Buchwald would hardly have built a house on the Vineyard if he thought he was going to be here all on his lonesome.

The consensus seems to be that the rat pack as presently constituted is led by Lillian Hellman, the left-wing authoress. Subsidiary members include CBS anchorman Walter Cronkite (who was recently turned away curtly by a head waiter for being *seven minutes* late for his reservation—a true story, everyone swears); opera singer Beverly Sills; James Reston of *The New York Times*; Mike Wallace of CBS, William Styron and John Updike, authors; and others too numerous to mention. Oh yes—Anthony Lewis, who, they say, roughs it in a tent when he comes. It all has to do not just with Getting Away From It All, but lettting everyone (who counts) know that you're doing just that. Be it reported, incidentally, that Tom Wicker doesn't come here. He would be more than welcome, of course. There's also someone called John Belushi (if you know who he is—I don't). Everyone has been saying that he just bought Robert McNamara's house on the Vineyard.

Somewhere in the middle of the island is a large house owned by a World Bank official—not McNamara but someone junior to him. I went to a big party there one evening. It must have cost a lot to give the party, let alone build the house. But then World Bank officials are engaged in such important work (ending poverty in the Third World) that they don't have to pay taxes. Two or three people told me that John Belushi had just bought McNamara's house. It was one of those parties where everyone felt awkward because no one knew anyone (except it was a good bet they were from Washington), but everyone had shown up promptly because (if the truth be told) on Martha's Vineyard there's not a whole lot to do. So pretty soon I, too, started to tell people I was introduced to: "Did you know that John Belushi bought Robert McNamara's house?" And of course they did.

The water is too cold to swim in. The island—larger than you might imagine—is overgrown with scrub, pine trees, small oaks, and masses of prickly vegetation of every description. Farmland has reverted to forest. Too much Ecology and not enough Energy, I reflected. But that is just the way the rat pack likes it—somewhat resembling an underdeveloped country, and itself meriting foreign aid or World Bank loans in due course.

One thinks back inevitably to the days of whaling—to those dangerous three- and four-year voyages. "Why did so many Vineyard men go whaling?" inquires Gale Huntington in a local guide book. "A man could make a living, and sometimes a good living, at sheep farming or fishing. Perhaps he could make a better living as an officer on a merchant vessel or as a pilot, but only by whaling was there a chance for a young man from the Vineyard to become really wealthy. . . . Many Vineyard whalers reached command while still in their early twenties."

Lest we forget, whaling was energy production. I suppose oil wildcatting would be the modern equivalent. But today's enervated upper classes do not aspire to anything more than energy conservation. The island seems to be alive with passive solar talk, windmill demonstrations, glass recycling projects. One day I went to the fair at West Tisbury, and one of the first booths I came to was occupied by the Energy Resource Group of Martha's Vineyard, "created in 1976 by a group of island residents concerned with the lack of accessible information about energy conservation and alternatives."

A girl in cutoff blue jeans handed me a bumper sticker: Support Vineyard Recycling. (I could put it next to my Vote for Anderson sticker.)

Nearby there was a pitiful little display of aluminum cans, bottle tops and whatnot. I wasn't quite sure what the point was. Maybe there was a right and a wrong way to throw cans away or scrunch them up or something. Such tiny horizons.

I asked the girl where the money came from.

"We're CETA funded," she said. (This was a government-funded "jobs" program in the Carter years.)

I guess energy conservation is one of the ways the upper classes go on welfare these days. The Energy Department or the Labor Department will come up with some money and it spares one the humiliation of going down to the local welfare office. Out of the question, of course, to go off and find work on one of those oil rigs . . . roughnecking . . . the modern-day whaling. That would take too much energy, which we all know is in short supply in America these days. Captain Ahab, where are you now that we need you?

* * *

Now that Senator Edward Kennedy has been knocked out of the presidential race, the Vineyard (summer visitor contingent) has switched over pretty solidly to John Anderson. The regular Vineyard residents, on the other hand, never were too enthusiastic about Kennedy. "They know too much," my host hinted, in reference to Chappaquiddick. I dutifully went sightseeing at the fateful spot and heard the following rat-pack explanation of the tragedy. Growing tired of all the earnest political conversation in the cottage that night, Mary Jo Kopechne slipped out for a snooze on the back seat of Kennedy's car. Kennedy then came out and drove off for the beach where he could enjoy a solitary, Kennedy-esque stroll before turning in . . . and really didn't know the young lady was in there with him when the car slid off the bridge. . . . (Just thought I'd pass it on.)

At one point I heard the year-round Vineyarders, who are said to be mostly for Ronald Reagan, described as the "townies." This was in connection with the Great McDonald's Hamburger Episode a few years back. The hamburger people, perceiving a good market, wanted to open a new outlet in Vineyard Haven. But Lillian Hellman thought otherwise. Bill Styron thought otherwise. I'm sure that Walter Cronkite thought otherwise. And unless I'm much mistaken Scotty Reston's *Vineyard Gazette*, with a cause to champion at last (values, y'know), thundered against the prospect of Big Macs on the Vineyard.

"The townies wanted it," a visitor recalled. But what the townies want and what the townies get can be two very different things, depending on the will of the gownies.

To get back to politics, briefly, it is worth considering this business of Kennedy supporters backing John Anderson for president. This suggests the rat pack is content to see Reagan elected in November. Why? The truth may be that there is so little cohesion at the center of U.S. political life at present that political goals can more easily be advanced in oppposition than in office.

One fears for a President Reagan, should he be elected. One can already hear the drumbeat of hostility from Congress, the wailing obbligato from the press, the crescendo of criticism—all building up to a contrived crisis. Better, perhaps, to be in opposition in this weak-willed, energy-conserving period of U.S. history.

From Martha's Vineyard I went to New York City, where I somehow managed to meet George Gilder at Grand Central Station. We went by train to Tarrytown, N.Y., whence we were driven to Pocantico Hills, the Rockefeller estate. George and his wife Nini were staying in one of the houses on the 3,500-acre estate, of which it has been said that God could have made it just as well if only He'd had the money. Another house was occupied by Henry Kissinger, another by Happy Rockefeller's chauffeur. Nelson Rockefeller himself was dead by this time. I met Happy by the pool, but she didn't recognize me from the San Souci. Inside the pool I met Henry Kissinger, doing a sedate breast stroke while his dog looked on faithfully from the sidelines. Harry Evans, the editor of the *Sunday Times* of London, sped back and forth at what seemed racing speed for several laps. I gather he was helping Kissinger with his memoirs.

Much of the estate was in the process of being deeded in some way to New York state, with the proviso that the Rockefeller family could stay on—for how long I don't know. I gather that John D. Rockefeller, the one who made all the money, mainly built the estate for his wife but himself spent little time there. I peered in through the windows of the big house—furnished but seemingly deserted. There was a well-kept golf course with scarcely a sandtrap. I played a lonesome round and saw no sign of anyone. In a deserted summer house I found hickory-shafted mashies and niblicks, perhaps left behind from a golfing party fifty summers earlier. Rather more prosaic (technology-aping) pieces of modern art were dotted about the grounds, having been collected by Nelson. They seemed deserted on the grassy slopes, now that their patron was gone.

The place I liked best was a well-kept barn the size of a small aircraft hangar, containing the ancient-to-modern lineage of Rockefeller automobiles. There was another barn for the antecedent carriages, but I never did find it. All these roadsters with running-boards, going back to the earliest models, were in perfect condition, their gas tanks topped up, keys in the ignition and ready to run. I climbed aboard a Packard tourer and pressed the starter. It coughed into life and soon was running smoothly. I was tempted, but decided against putting it in gear. So I just sat in the driver's seat for a few minutes, like a child, with my hands on the steering wheel.

A side door opened and an elderly but well preserved gentleman walked into the barn, with a lady I took to be his wife. He looked over at me and smiled but said nothing. George Gilder, who was just then finishing *Wealth and Poverty*, told me later that this was Laurence Rockefeller, brother of Nelson, David, Winthrop and John D. III; and his wife Mary. They drove off in a modern, perfectly plain Chevrolet.

IPS BIRTHDAY PARTY

I wondered what I should wear for the Institute for Policy Studies' 20th anniversary party at the Pension Building. On the invitation it said, "Attire: picnic casual." And something about blue-grass music. I tried to imagine Richard Barnet and Marcus Raskin twanging mandolins and hopping about in cowboy suits but it was too difficult. I already knew exactly what they'd be wearing: tweedy sports jackets, grey slacks (dark grey), preppy button-down shirts and nondescript ties. In other words—media rig. So I dressed the same, with the risque exception of navy blue corduroys (1930s Hampstead-socialist, poetry-reading, pipe-puffing type). Also bright red belt, to tone in with mood of the evening. IPS was founded by Barnet and Raskin in 1963. It is described by Charlotte Curtis of *The New York Times* as an "independent research organization."

I paid my tax-deductible $35 and entered the Pension Building, a great rosy brick structure from the Robber Baron Era and one of those buildings that brings out the sentimental side of left-wing intellectuals. They think of these old buildings, with their adorable quoins and corbels, as their very own cathedrals, and in their secular way they moon about inside them, communing reverently with the Commodore Vanderbilt trappings, the Frick friezes, the Jay Gould joists. And so it was to be tonight at the Pension Building, vintage 1883, designed by Gen. Montgomery Meigs and the whole containting 15 million yummy red bricks.

Hundreds of lefties were pouring in at the main entrance, all kissing one another on sight and swearing fealty to Mark and Dick. Round tables were decorated with red and white balloons (true-blue missing, get it?). A band was setting up amidst the usual electronic shambles, the guitarist twanging exploratory notes like arrows out across the football field-sized space.

Here was Abner Mikva, the U.S. Court of Appeals judge (called a moderate by our confusing media taxonomists); over yonder Roger Wilkins, the Negro celeb. Here in earnest confab were Victor Navasky of the *Nation*, and good old sturdy left-wing philanthropist Philip Stern, whose foundation-matured Sears Roe-bucks have been bankrolling IPS since its inception. Richard Cohen of *The Washington Post* was on hand, as was his wife Barbara, the news director of

National Public Radio (later she moved to NBC). Cohen told me that he was "celebrating the rejuvenation of the Left." He is part of the *Post* brood that goes cluck cluck cluck at Mr. Reagan every day, and, in the spirit of detente, always gives the benefit of the doubt to Andropov, Arbatov & Co.

And here, finally, was Dick Barnet: thatchy-jacketed and professorial in manner, courteous as always, watchful, alert. Semitic in looks, he foresook Judaism in the 1950s and joined an independent, non-dogmatic Christian sect which celebrates fellowship at a storefront church on Columbia Road, N.W. I think Barnet may have been one of the first to recognize that the future of socialism lay with religion, not (as it had seemed in the 19th century, and the first half of the 20th) with science.

It had been several years since I had seen Barnet, but he was as affable and composed as ever. Barnet is a major figure on the American left, but no one I speak to seems to know very much about him. He writes books that in recent years have been serialized in *The New Yorker*. But he tells us nothing about himself, and I don't think a profile of him has been published anywhere. At this anniversary gathering, surrounded as he was by old friends many of whom he had no doubt not seen in ages, he was so attentive and undistracted as I spoke to him that I mentally took my hat off to him. He mentioned his new book—something about the alliance: it too would appear in *The New Yorker*. He said he didn't read the magazine much but he wanted me to know that he had rather enjoyed William F. Buckley's recent contributions to it.

The journalist Richard Grenier told me that in 1963 he met Barnet at a Communist-dominated meeting in Paris—World Jurists For Peace, or some such title. Senior Communist Party officials from the Warsaw Pact countries were among the delegates: "either the Minister of Justice or the Deputy Minister," Grenier recalled. Chief delegates from the West European countries were members of the central committees of their respective Communist Parties. The American representative was Richard Barnet. Grenier at the time thought it was incautious of Barnet to be so openly associated with the Communist world. At that time "an endorsement from Moscow was still considered death," Grenier said. But Barnet saw the future more accurately than Grenier. "I greatly underestimated the influence these people would have in the coming decades," said Grenier. He had an opportunity to speak to Barnet after the meeting. Barnet made small talk, saying he was going into the think-tank business: IPS was founded at about that time. Barnet said that think tanks were the wave of the future and would, for example, create opportunities for those with brains but not able to make partner in the big law firms.

I myself had met Barnet in the spring of 1976, at a dinner party given by the number two man in the Pakistan Embassy, Iqbal Riza. This was at a time when those fashionably on the left in Washington were jetting off to Cuba, mostly in the entourage of Senators James Abourezk or George McGovern. The Sandinistas hadn't yet come to power so Cuba was still the "in" place. The visitors would lie on the beach and then wait excitedly in their hotel rooms for the midnight call from the nocturnal Fidel. On their return to Washington they would greet old friends who had been left behind with little moaning noises of ecstasy: Cuba had been *that* good.

There had been yet another jaunt to Havana, and I recall that there were some moaned reunions that night, before we sat down to Iqbal's excellent curry. At one end of the table was Tad Szulc. I sat opposite a not-very-nice Chilean Communist named Orlando Letelier, who outdid all the others in the stridency of his anti-American table talk. A few months later he was blown up in an automobile on Massachusetts Avenue, a couple of hundred yards from where I had been sleeping until the blast woke me up. Next to Letelier at the dinner party sat Richard Barnet. No stridency there. No sentimental Cuba-moaning from him, either. He was cool, attentive, watchful, reasonable—hard to beat in an argument.

Now I see that the Institute for Policy Studies is co-sponsoring a Minneapolis disarmament conference in May, 1983, along with Georgi Arbatov's Moscow-based Institute of the U.S.A. and Canada. Here is another bold move by IPS. Arbatov is a close friend of Yuri Andropov, and a close student of the United States (see his book *The Soviet Viewpoint*, just published by Dodd, Mead.) But I stress the word student, not instructor.

Back at the IPS party, we all had our spareribs and chicken, and after a hearty round of applause for the honored guests from the Nicaraguan Embassy and the Cuban interest section, an effusive welcome from the mayor of Washington, and a resolution declaring April 5 "IPS Day" in the District of Columbia, (students of left-linkage might find it rewarding to look into links between IPS and the DC government), we sat back for a "roast" of Barnet and Raskin.

Such well known humorists as George McGovern, Cora Weiss, Ron

Dellums, Robert Kastenmeier and Ralph Nader took turns on the dais. We also heard from Harry Belafonte, who is far more to the left than I had realized. McGovern, who made over a million dollars trading real estate in the decade following his 1972 presidential race, commented that he had given the matter some thought that afternoon, but he "couldn't recall any indication at all from Mark or Dick that they ever wanted to be either red or dead." Moreover, McGovern added, he had "always had a certain appreciation for people who think that avoiding death as long as reasonably possible is a fairly high priority. . . . The color I want least for my grandchildren is dead."

Paul Warnke was master of ceremonies, rather to my surprise. He gave a speech toward the end of the evening. One wonders what can have been going on in his head since he was an assistant secretary of defense in the late '60s. He gave me the impression that there was one big thing he had learned from the Vietnam debacle—America is on the losing side and we should reconcile ourselves to this prospect as speedily as possible. Then, under Jimmy Carter, he headed up the Arms Control and Disarmament Agency for a couple of years before returning to his law practice, shared with Clark Clifford.

Warnke is a choleric-looking fellow. You get the feeling he wants to show the world he's made of sterner stuff than his right-wing critics implied at the time of his confirmation hearings. Perhaps that was why he was willing to host the IPS show. He said, I think misleadingly, that "unconventional thought" is "still somewhat suspect" in Washington, and that "radicals of the right have managed to escape this stigma, and to retain social acceptability."

This he had exactly the wrong way 'round, I thought. Here was a party that journalists were attending not because they were covering it but because they knew it was their social and intellectual milieu: Their friends were there and they wanted to be there too. If Warnke wants the unfashionableness of his friends to demonstrate the unconventionality of his thought there are a few very untrendy suburban get-togethers in Arlington and Fairfax counties I could direct him to. He would no more relish the conservatives' jug wine than he would their opinions. Even with their man in the White House, conservatives have remained social outsiders in Washington.

It was unnecessary, and probably also tactless of Warnke to bring up the subject, hinted at by Robert Moss and Arnaud de Borchgrave in their novel *The Spike*, of KGB influence over IPS.

"Those who would call into question whether American might

always makes right can find themselves the subject of such things as a cheap novel that suggests KGB control of ideas," Warnke told the 800 assembled. "Can you imagine Mark and Dick confronted with the KGB? Andropov would have applied for a transfer! No, IPS is uncontrollable. It can't be controlled by the CIA, or by the FBI, or by the KGB. It's a demonstration of the fundamental independence of American thought."

This was slightly inaccurate in that IPS thought bears a strong resemblance to socialist thought the world over. But still, Warnke was right in his main contention. Mark and Dick are not acting on anyone's instructions. If anything, the direction of influence has been the reverse. The whole style of Arbatov's book much more closely resembles *The New York Times* editorial page or the IPS study than it does the speeches of Lenin. Haynes Johnson of *The Washington Post* recently noted Arbatov's "disturbing eloquence." Come to think of it, Arbatov more and more sounds like Haynes Johnson himself.

"The Soviets consciously borrow from the United States— technology, managerial know-how, style," Barnet wrote in his book *The Giants*. "But Americans, even those who would like to see fundamental change in the system, have little interest in the Soviet model."

Dick Barnet's sense of the west-to-east direction of influence is correct, to give credit where it's due.

Eric Hoffer in San Francisco

I met Eric Hoffer in the lobby of the Raphael Hotel in October, 1980, just three weeks before President Reagan's election. He was sitting quietly in a corner armchair, looking out of place in his laborer's clothes. A flat workingman's cap was perched on his head. In his knobbly hands he clasped a knobbly walking stick. Earlier we had corresponded. I had been flattered to receive his letters, written in his artless, looping hand. "Come any time," he had written. "We shall eat, drink and talk."

We walked down the street to a coffee shop at the corner of Mason and Geary. He didn't exactly tap the ground with his stick but it was obvious that he saw only dimly. He spoke in a strong Bavarian accent. "It grows thicker as I grow older," he said. A couple walking by recognized him and stopped to acknowledge him and exchange a few words. Hoffer was obviously pleased to be recognized and to find that he still had some fans. Later on he told me how much as a writer he had wanted and needed to be praised.

I was unprepared for the contrast between his literary and his personal style. "I am a vehement person, a passionate person," he said. "But when I write, I sublimate. It's not natural for a passionate person like myself to write the way I write. I rewrite a hundred times, sometimes, so that it is moderate, controlled, sober. I need time to revise, time to change."

In his first book Eric Hoffer had written about "the true believer." Now I could see that he was himself part true believer, but that he had contrived a judicious disguise for himself in his books. He was in truth a "fanatic," he told me as we walked along. And his great cause was one rarely articulated by intellectuals in America: America itself.

"Do you know," he told me as we sat down, "I was never accepted by the San Francisco literary establishment. It was because I have praised America extravagantly." Herb Caen, the gossip columnist of the San

Francisco *Chronicle,* particularly seemed to despise him, said Hoffer. He said of Caen that he had talent but that he had frittered it away. It seemed not to bother Hoffer that he had failed to win the affection of the local literati. On this he was detached, savoring the irony. We ordered coffee. As we waited, he took issue with something I had written to the effect that America might now be in decline. He thought this was premature.

"America is a fabulous country," he said loudly. Heads at the lunch counter began to crane around in our direction. "It's not so conspicuous because the jetsam and the dirt are all on the surface. I remember when I first started to think about writing a book. I was going to go through my life and write down all the kindnesses done, right from the beginning. It is a really fabulous country. Consider the lengths people will go to to come here. And who built this country? Really nobodies. Nobodies. Tramps."

I said that I worried that an alliance of intellectuals had finally learned how to manipulate government to achieve the power they craved. They were slowly getting the grip that had so long eluded them.

"Well," he replied. "We'll see what Reagan does." He spoke highly of Reagan, saying that he had always been underestimated, just like America itself. "It is easy to underestimate America," he said. "We underestimate it, our friends underestimate it, our enemies thank God underestimate it. But somehow there is a tremendous vigor in this country. It is true that our intellectuals are becoming much more influential. They are shaping public opinion and so on. But that won't last. There will be a reaction against it. For years I wondered how and when the silent majority is going to wake up. And I didn't take the factor of religion into account. That's a beautiful situation right now. [Hoffer was referring here to the rise of the Moral Majority.] You see, there used to be a time when you had great leaders everywhere: here was de Gaulle, here was Churchill, here was Adenauer. Then all of a sudden there wasn't a great man on this planet. How come? What happened?"

He took out a pack of cigarettes and lit up, saying that he wasn't supposed to smoke but that he would anyway. "You see," he said, "you are going to have great leaders if it is possible to have power. But when the sources of power are so inconspicuous and so hard to manipulate, then leaders do not appear. Right now religion is the only source of power that's there. You have Pope John Paul, you have the Ayatollah Khomeini, you have them popping all over. And Reagan too is now tapping a religious strain."

He told me that he was working on a book he would call *Conversations*

With Quotations. "If you ever come to my room you'll see. I have just an enormous number of cards. All my life I used to write down anything that I thought I wanted to remember. So I've got a thousand or maybe two thousand quotations. And every time, after that quotation, I am going to talk to that man."

It was soon time for Eric's long-time friend Lili Osborne to pick us up in her car. We would be going to a restaurant opposite the apartment building where he lived, overlooking the San Francisco docks. Lili turned out to be a motherly woman of Italian background who taught school in Redwood City. She had known Eric for about thirty years. Eric sat beside her, and as we drove along toward Market Street he pointed out a Unitarian Church as though it were a regular landmark on Eric Hoffer's Guided Tour of San Francisco. "That's where all the radicals go who have lost faith in radicalism," he said. Then for some reason the Pope once again crossed his mind. If he ever had a chance to meet the Pope he would tell him: "Go slow! The church has lasted so long. It has discovered the secret of survival."

As we drove toward the waterfront he said: "We forgot all about the human condition. We forgot what evil is. With Burke there is still a whiff of evil, in *Reflections on the French Revolution.* But with Freud, all we need is a little screwing. The man was a pervert I tell you! He admitted that he brainwashed little girls into saying that they wanted to play with him. He wants to infect us with his own sickness and then offer psycho-analysis as a cure."

When we entered the restaurant Hoffer put on a great display of bonhomie: the gourmand come to his feast. "Bon appetit!" But I won-dered if it wasn't put on for show. Normally, surely, he didn't bother too much about material comforts. He was greatly amused because the restaurant, a fashionable one, was externally "French" but actually owned and run by Chinese—a people he much admired and wished would come to America in greater numbers.

The conversation soon turned to a recent CBS documentary about the rising political power of homosexuals in San Francisco. The producer, George Crile, had interviewed Hoffer for the program but had not used the footage, much to the relief of Lili, who was worried about the hostility his remarks surely would have evoked. What concerned Hoffer (as he apparently had said on camera) was the increasing shamelessness and militancy of the homosexuals.

"There cannot be civilized living without shame," he said, as slim waiters flitted by. "Shame means the acceptance of rules that cannot be enforced by coercion. If you a have shameless society you haven't got a society at all. In Russia, for instance, you accuse your father, your mother. Everybody accuses his friend. There is no shame, and no society. All you've got is a comglomerate of people held together by coercion. Hesiod has a beautiful saying: When the Goddess of Shame will depart, our society will fall apart. This is what bothered me about homosexuality: they were not ashamed to admit it. And they want to convert people to their oun thinking." (He pronounced "own" to rhyme with "noun.")

"Something different has happened in . . ." Lili began.

"All of a sudden our values are breaking down," Hoffer said. "Our norms are breaking down, and the rats are coming up from everywhere. They are a symptom of disintegration and decadence. I don't know if we can reverse it. The question always comes now, can we reverse things? It's an uphill battle. Of course, they always fall back on the Greeks. But the Greeks were a different kind of homosexual. There is a tremendous violence going hand in hand with the homosexuality in San Francisco. There is violence in the act there. They are hurting each other . . ."

The waiter at Hoffer's elbow was ready with his menu recital. "As an entree tonight I have angler . . ."

"Nothing is too good for us, remember that," said Hoffer. "I am floating through the air with the greatest of ease. It is good to be alive. Count your blessings." The last phrase was a favorite of his. He repeated it several times in my presence.

Eric Hoffer was born in the Bronx, supposedly in July, 1902, but he may really have been born a year or two earlier. There are hints in his writings and interviews that he may have changed his age, just as it is possible that some of the events of his early life, told and retold to interviewers, were imagined. The title of his posthumous memoir, *Truth Imagined*, suggests this. Lili Osborne also believes that Eric may have been born in Europe as they were never able to locate his birth certificate.

Eric told me that his father was a "small town atheist from Alsace." His parents came to America around 1900. "My father was a self-educated cabinetmaker," Hoffer wrote in *Truth Imagined*. "He had nearly a hundred books, in English and German, on philosophy, mathematics, botany, chemistry, music and travel. I spent passionate hours sorting the books according to size, thickness and the color of their covers. I also learned to distinguish between English and German books. Eventually I learned to sort according to contents. I might say, therefore, that I had learned to read both English and German before I was five."

When he was five (according to the oft-told story) his mother fell down a flight of stairs while carrying him. Two years later she died and he went blind. He recalled his father referring to him as the "blind idiot" or the "idiot child." He attended no school, recovered his sight when he was 15, and then began to read all he could lay his hands on (beginning with Dostoevsky's *The Idiot*, which seemed as though it might have been written with him in mind). His father died in 1920, whereupon Hoffer set off for California. He proceeded to go through life "like a tourist," he said, because he had been told that, like his father, he would not live to be 50. He stayed in California for the rest of his life, very rarely leaving the state. A few uncomfortable minutes in a Mexican border town was his only trip outside the United States.

While doing part-time work as an agricultural worker and gold prospector in California, Hoffer discovered Montaigne's *Essays* in a public library. The thickness of the volume reassured him; it would last him through the season. He learned whole passages by heart and would recite them to his fellow hoboes in the San Joaquin Valley. "And then, every time something came up, like money or women, they would say, 'What does Montaigne say?' And I would quote him. There must be hundreds of hoboes going up and down San Joaquin Valley quoting Montaigne."

Hoffer read the Bible for the first time when he was 27, and this had an even greater influence on him than Montaigne. The Jewish nation and its role in history preoccupied him for the rest of his life, even though he wrote about the subject hardly at all. God, Hoffer said, was simply an invention of the Jews. But at table he usually asked someone to say grace. When the theory of evolution came up for discussion, Hoffer said that he didn't believe it at all. "It is easier to believe in God," he said.

With the outbreak of World War II, Hoffer found steady employment with the Longshoreman's Union in San Francisco, headed by Harry Bridges. Hoffer began to fill small notebooks, slowly piecing together *The True Believer*. He sent his only copy of the manuscript, in longhand, to Harper & Row, receiving a telegram of acceptance several months later. When I asked him if he wasn't afraid that the manuscript might have been lost he said no—by then he knew it by heart. The book was published in 1951.

It was a *tour de force*, favorably reviewed by Bertrand Russell. Hoffer later became immensely antagonistic toward intellectuals, but in most respects he was one himself. Arthur Schlesinger Jr., praised *The True Believer*, and Richard Rovere said in *The New Yorker* that it was the work of a "born generalizer." In an approximate one-sentence synopsis of *The

True Believer Hoffer wrote: "Faith in a holy cause is to a considerable extent a substitute for the lost faith in ourselves."

A later work, *Before the Sabbath* (1978), first attracted me to Hoffer. Here current events provided the stimulus for his thoughts about life and history. It is in the movement from the particular to the general that we find Hoffer at his best. (A possible criticism of *The True Believer* is that it is too inexorably general and too infrequently particular.) In *Before the Sabbath* he returned again and again to the anti-Americanism of the intelligentsia, attributing it to "leprous vanity . . . pretentious nonentities wanting to avenge themselves for being ignored." I wanted to meet him when I read that, and so I wrote to him, receiving his reply within a week: "Come any time . . ."

The day after we had dinner he invited me to his apartment. He lived alone in a tenth-floor efficiency in a building with a doorman. His apartment overlooked the waterfront and San Francisco Bay. It was, as one might have guessed, spartan but sufficient for his needs: a bunk bed, a plain table, two chairs, a small bookcase containing only a few favorites: Dostoyevsky, Montaigne, his own books (with some foreign translations of *The True Believer*). He kept up with *The New York Times*, the *Jerusalem Post* (weekly edition), *Commentary*, *The Economist*, and several other publications, but recently he had undergone an eye operation and he was beginning to find reading difficult. The view from his balcony brought back memories for Eric: he said he could smell his own sweat. At night, he said, he would sometimes dream he was back working on the docks, and then his muscles would ache when he woke up. That's why longshoremen are paid pensions, he said.

But he was not in a reminiscing mood today. He was holding one of his index cards, and he was trying to fathom, as though for the hundredth time, the quotation written on it.

> America is the most aggressive power in the world, the greatest threat to peace, to national self-determination, and to international cooperation. What America needs is not dissent but denazification.
>
> —*Noam Chomsky*

"Now," he said. "What do you do? This was during the heat of the Vietnam thing, I suppose. What do you do? You try to understand why they say these things. What do you think, Tom? You know them better than I do. You have been rubbing your brain against them. What is it that makes a man who is highly intelligent say such a thing? They call him a metaphysical grammarian. He was invited by Oxford to lecture. He couldn't be just a fraud. He's a very successful, prospering intellectual. What I know about his past is that he grew up in a highly orthodox Jewish household. Indeed, he says that all his ideas about grammar emerged from his familiarity with Hebrew grammar. He was born in this country. Somebody told me that he is good-looking. I've been asking people about him, you know. Not only does he side with our enemies, he sides with the enemies of Israel: Arabs, Palestinians, dissenters in Israel. I am trying to say something reasonable, and what I probably will say will be this:

"Chomsky loves power. He is also convinced of his superiority over any politician or businessman alive in the United States. He sees the world being run by inferior people, by people who make money, by people without principle or ideology. He thinks that capitalism is for low-brows, you see, and that intelligent people should have a superior form of socialism, and so anybody who interferes with this program is a criminal."

Eric Hoffer went to sit on his bed. "Tom," he said, "what gives people like him the confidence that they really know everything, that they are superior to everybody else? Just knowledge doesn't give confidence. If you went to school and you looked at your professors you would see that the brighter they were the less confident they were that they knew everything. There's something *else* here, something else . . ."

Then, reflecting on the amazing ungratefulness of some who have done well in America, Hoffer said: "Gratefulness is not a natural thing. There are two sorts of people coming to America. Both were nothing before they came, and both made good. One will say, I came as a barefoot boy, and look where I am now. What a good country it is! The other will say, I came as a barefoot boy, and look where I am now. What a bunch of idiots they must be! He sees his own rise as evidence of others' inferiority.

"There was a prototype for all this in Vienna," he went on, "toward the end of the Hapsburg Empire, before World War I. There was a group of very brilliant people, Jews and non-Jews, who were just glorying in the approaching doom. They knew that the end of the world was just around the corner. And I couldn't figure out how intelligent people, who liked to eat the good food of Vienna and sleep with the beautiful women of Vienna, should derive such tremendous pleasure from contemplating the approaching doom. And the answer is the same: they were so convinced that they were the fittest men to run the world that they wanted with all their hearts to wipe the floor and start from scratch. And they would show us how the world should be run."

Hoffer put the card down and said: "Well, I don't know. Maybe I should throw out Chomsky and not have any business with the son of a bitch. But he needs explaining. How come this hostility on the one hand, and this confidence on the other? Imagine what a disaster it was when we had the Depression and Roosevelt, that crippled aristocrat, got into power. He had a whole bunch of these chomskies and bomskies around him, telling him how to do things. These were people who didn't have the least idea what this country was all about. They knew that it wasn't built right, they called our Constitution a horse and buggy thing, and so everything had to be reformed. What a disaster that these intellectual swine had such a free hand! Roosevelt is a watershed in American history. The time will come when American history will be divided into B.R. and A.R. Before Roosevelt, if something went wrong, you blamed yourself. After Roosevelt, you blamed the system—everybody but yourself."

Somehow Karl Marx came up. "All Marx's predictions turned out to be wrong," Hoffer said. "All his doctrines turned out to be wrong. And yet these wrong predictions and doctrines played a more fateful role in shaping events than anybody else's. Because all the slogans of recent revolutions came from Marx's wrong predictions and wrong doctrines." Here Hoffer emitted an unexpected cackle of laughter, followed by a great and mysterious groan. "That was Marx's greatness. And of course the reason Marx had such great appeal was that he really hated humanity. Oh, that's their man, you see. You hate your own country, you hate your own species. You know Douglas, Justice Douglas? He said 'I finally came around. I am on the side of the fishes against the fisher.' That's his great achievement, in his old age. Nature is pure, it's healthy, salubrious. Man taints, pollutes the world. Oh, Tom. Get the book. See what Calvin says about humanity."

I read from his card index quoting Calvin: "The first man and woman brought sin into the world, and immersed all posterity in the most terrible pestilence, blunders, weakness, filthiness, emptiness and injustice."

Then I read Hoffer's written reply: "Is it possible to have ardent faith, even in man, without a loathing for man as he is? To a fanatically religious person the only sinful living being is a polluter of the universe."

Hoffer thought about this for a moment and added this coda: "To

Chomsky, Kunstler, Douglas and all the others we are polluters. The moment we get in a forest we pollute with our breath, we pollute everything. They hate humanity. Humanity stinks in their nostrils. And they are our saviors, you see."

He suggested that I pick another card from the file, directing me to Arthur Bryant, the English historian. (Hoffer's cards were only indexed up to C.) He told me to read it aloud for his benefit: "There has never been a time in history when the Jews have not been news. And the periods during which the Jews have occupied and dominated Palestine have been the most exciting and significant in man's sojourn on this planet."

At this Eric gave a little sigh of pleasure. "Here you have a Wasp, an Anglo-Saxon, with not a drop of Jewish blood in him, saying something that no Jewish chauvinist would dare say!" He let out an odd, whinnying peal of laughter. "This is something I *can't* explain," he said, puzzled and serious once again. "The uniqueness of the Jews. I can describe their uniqueness, but why they became what they are I don't know. It's the greatest mystery in the world for me, and I have been preoccupied with the Jews since 1929. It's a special thing to be a Jew, and this is what most Jews don't know. They think they are like others, but they are not."

Hoffer said he had been invited to Israel many times but he didn't want to go because he was afraid that he would be "disappointed by reality." He said he knew the country "the way I know the palm of my hand. I can describe how you land in Jaffa or Haifa and how you go to Jerusalem. I can describe the road exactly, as if I was there. Of course, I have seen photographs."

He fell silent for a while and seemed to be quite tired. He lay down on his bed. "The Jews have the atomic bomb," he said, "and they're going to use it if cornered." He opened his eyes and added: "I don't think Egypt will attack Israel there. The only place I . . . I think Russia will somehow stumble into it. If, let's say, Syria starts something, and the Israelis get in there and wipe the floor with Syria, then the Russians might send an army in and start to order the Israelis around. And the Israelis will say: What the hell. Let's see who blows up whom."

In the last year of his life, according to Lili Osborne, Hoffer became increasingly preoccupied with the Middle East, especially after one or two of his suppositions came true. In the course of a CBS television interview with Eric Sevareid in 1967, Hoffer remarked that "If Israel fails then history can have no meaning."

When I first wrote to Hoffer, in 1979, I asked him something about the Soviet Union and its prospects. He replied as follows: "As to Russia: She will not change so long as the Occident remains dynamic. Only when the West declines and becomes negligible will Russia venture to experiment with democratic socialism or whatever. She may then even cast herself in the role of a modern Occidental power facing the primeval Chinese dragon."

Six months later I again went to see Eric Hoffer in his apartment. His health had noticeably deteriorated. He said he wanted me to entertain him a little, to read him a few pages from *The Idiot*. When he asked me if I had read the book and I said no, he seemed disappointed in me. What were they saying about Reagan in Washington? he asked. He groaned when I said "simplistic." He lay on his bed with his eyes closed. I thought he was asleep when the buzzer rang and a voice announced that an old friend was downstairs: Selden Osborne.

Eric came to life immediately, sat up, and told me that here indeed was an interesting fellow—a "true believer," he said. They had known one another for many years. *The* true believer? Eric implied that Selden had contributed to a composite portrait. He and Selden had worked together as longshoremen on the waterfront. "He's a doctrinaire socialist," said Eric, just before Selden came into the room.

Selden was perhaps ten years younger than Eric. Slim and well preserved, with a trim beard, he was wearing a light khaki safari outfit.

"I was an active participant in the union," Selden told me. "Eric was more of an observer. He always said the table was laid for him."

"I appreciated it," said Eric.

"Eric was violently anti-Communist, but he gave credit to Harry Bridges," said Selden. "Bridges was afraid of educated longshoremen."

"No, he was afraid of longshoremen he couldn't use," said Eric, who was by now very much himself again.

"Eric always said that America is the country of the common man," said Selden. "I disagree. It's the most imperialistic country of all. Also the union under Bridges. I favored democracy in the union. Eric said it wasn't so important. Do you remember that, Eric? You said you never grew up with a sense of democracy because you never went to school."

Eric was leaning forward on his bed, and through the open window you could see the sailboats in San Francisco Bay leaning over in parallel. "Repeat what you said," he demanded. It was as though the years and decades had fallen away and they were back together, arguing on the waterfront.

Selden repeated it and added: "I think that the leadership of this country is hellbent to rule the world."

"His antagonism to this country is because it doesn't give him a role in leadership," Eric explained. "It's very difficult to become a leader. . . ."

"Unless you have a lot of money."

"Noooooooo-o-o-o. Bridges didn't have no money. You're not a democrat, Selden. You're an exclusivist. You resent any setup where an ignorant son of a bitch can run for office and get elected."

Then Eric lay down on his bed, coughing.

"Take Truman," he said, after a while. "He was an ignoramus and yet he made a good President. To me these things are all miracles." He challenged Selden to name a better country.

"Better country? I couldn't say. I might have said Australia or New Zealand. There is some advantage to a parliamentary form of government."

"I learned from Selden all the time," said Eric. "I classified him as a fanatic." He added that fanatics were often important and that "if you have a real cause that needs fighting you had better get some fanatics because you won't win without them."

"In those days I was a dissident Trotskyite," said Selden. "I remember I once took Eric to hear Max Schachtman. Most of the educated longshoremen were followers of Bridges. Bridges was a Communist and Eric an anti-Communist. The union paper never acknowledges his existence, even after *The True Believer* came out."

"I never spoke a word to Bridges," said Eric from his bed. Finally he sat up and said: "I don't want to be nitpicking, but in my case it was not Bridges but America that set the table for me. I asked Lili to put on my tombstone: 'The good that came to him was undeserved.' Many times I had ignorant people correct me. I never found a common man who would agree with Henry Adams that America was created by a bunch of crooks. It was always axiomatic for the common man: America was the last stop. If you couldn't make it here, you couldn't make it nowhere. It's a sort of treason to complain about America. Selden's disagreement was crucial. But I love him."

I left with Selden. As we walked toward the center of San Francisco he told me: "My whole concern today is disarmament." There was a place where some friends of his lived, about fifty miles down the coast, set among the redwoods. An old property with a goodly acreage. He would be going there in a few days, and together they would have some good discussions about disarmament. He had graduated from Stanford University and later had shared an apartment with Clark Kerr in Palo Alto. He had joined the longshoreman's union because he believed at that time that the revolution would originate with the working class. But he now knew that this had been a mistaken belief. As for Eric Hoffer, Selden said: "All his conclusions are wrong—every one of them. But he writes beautifully and he asks the right questions."

That evening I was in the car again with Eric and Lili. She was driving him back to his apartment, where she would make up his lunch for the next day, and then leave him once again to contemplate his index cards in his customary and preferred solitude. Eric mentioned the hopeless impasse of all his arguments with Selden. Neither one could ever persuade the other of anything, and so it had been for over 30 years.

"Selden is so anti-American that it frightens you," he said. "Born and raised in America. If anyone is for America it should be him."

There was a silence, and Eric turned to me and asked: "Do you think he likes me—deep down?"

I hesitated. Lili Osborne had been married to Selden. Then, some years later she divorced him. It was Selden who had introduced her to Eric Hoffer at about the time that *The True Believer* was published.

Lili turned the wheel of the car, and changed the topic of the conversation. She said that Selden was a man of strong convictions. "He has gone to jail for what he believes."

"Of course, ready to sacrifice himself for the cause," said Eric, gripping his stick once again between his knees. On his head was the same old cloth cap that he seemed to wear everywhere.

"Not many people are willing to die for what they believe."

"Lili, I have told you a hundred times. It is easy to die for what you believe. What is hard is to live for what you believe."

Beyond the Ochre and the Umber

THE STERNWOOD PLACE ON ALTA BREA

It's surprising how few passengers bother to look out of the airplane window. You would think they might find it worth a glance, at least. But they are too busy studying real estate maps of Los Angeles, lawyers' briefs or wearisome government reports. I'm studying a large-scale map myself. Colorado is spread out below me, partly obscured by the wing of the DC-10. The unseen contrail behind us is plowing a resolute furrow below, across ochre and umber squares, suede-colored rectangles. Soon the property borders cease entirely and our aerial shadow unflinchingly advances across papier-mache foothills, with fir trees clustered on pinnacles like iron filings attracted by some invisible current. To the north there is a great white ice-sheet. Here below purpled hills, rilles and Rockies; crumpled crevasses, dry-branching gullies, the wrinkled earth's ancient hide, the Colorado River, the West! Sandstone monuments, salmon-pink desert . . . odd tracks from nowhere to nowhere, irrigated semi-circles, another hour of desert.

Eventually one began to see irrigated land, the occasional reservoir, outposts of civilization, the Los Angeles Aqueduct bringing water from the Owens Valley; inhabited outskirts; reclaimed desert; San Bernadino; platoons, regiments, armies of dolls houses in parade formation; and without warning, the Pacific Ocean. Within a few minutes we had landed at Los Angeles International Airport, or LAX as everyone seems to call it.

It had all been arranged by telephone. For the next two months I would be "writer-in-residence" at the Los Angeles *Herald Examiner*, one of the Hearst newspapers, and I would be writing a column three times a week.

On the airport concourse, minibuses hurried about and a stiff breeze stirred the palm trees. It all seemed strangely familiar, and then I recalled Jeremy Pordage's arrival in Los Angeles in Aldous Huxley's *After Many a Summer Died the Swan*. Huxley was immediately intrigued by the signs: EATS COCKTAILS JUMBO MALTS OPEN NITES FACIALS. At that time the city was at the leading edge of contemporary Western culture. I wondered if it still was. You wouldn't guess so from the immediately visible signs: TRAVELODGE AVIS HOWARD JOHNSON. What *are* the indicators of the cultural frontier? Are they still in English? Or have they been miniaturized into microprocessors? Or are they in Japanese?

For enlightenment on such matters I telephoned Ben Stein, a former White House speechwriter, investigator of Hollywood and collector of facts and figures. His book, *The View from Sunset Boulevard*, documents the prevailing anti-capitalist trend in the film industry. He suggested that I meet him at The Palm restaurant on Santa Monica Boulevard, "a very famous street in Los Angeles." I should have said that I had already been down that street with Philip Marlowe. The ambience of Los Angeles, 40 years after Raymond Chandler's detective was in his prime, kept bringing back Marlowe. But I said nothing.

"You'll recognize the restaurant," Stein said, "because there will be about a million Mercedes and Rolls Royces outside." He spoke in a slow, droll voice. He had recently sold a script to the movies—quite a feat since (he said) one in a thousand scripts is sold, literally, and of those that are sold one in 250 is actually made into a movie. Four chances in a million! As Stein said, many of those who struggle against such odds conclude after a few years' failure that "the system" is to blame: They are unfairly at the mercy of crass businessmen. So as they sit at their typewriters they are predisposed to take it out on capitalism.

On the freeway there were signs illuminated with terse Caltrans homilies: SMILE EVERY MILE TRY RIDESHARING. I had time to drive past the Sternwood Place on Alta Brea, familiar to Marlowe fans, and was glad, somehow, to see the maroon Packard still in the driveway, complete with chauffeur in shiny black leggings.

Stein was waiting in the Palm, sitting alone in the center of the restaurant (dark, cool, sawdust on floor). His immediate topic was money, which he said was uppermost on everyone's mind in Hollywood. I gathered that one of the great problems was that people never know when they are well off. "I had lunch in this very restaurant with a 16-year-old actress last week," he said. "She's making $8,500 a week. But she says she's being cheated."

Here he paused to wave at a record producer, and he discreetly pointed out a very thin TV actress who has "a $400-a-day cocaine habit." Then, he said, "take the writer who is paid $200,000 for a script. He thinks *he's* being cheated. He will say to himself: 'Look at Norman Lear. He gets $20 million.' Or take another writer, and let's say he was paid $300,000 for a film script. The film was a hit—grossing $40 million. But he has no 'points.' No share of the profits. So he thinks he's being cheated too. 'There would be no picture without my script,' he says to himself. And then take Norman Lear. He gets $20 million, let's say, but he says, 'Look, the network is making $100 million.' And because of taxes, Norman only keeps a few cents on the dollar anyway. So everyone thinks he is being cheated."

So there you have a kind of pyramid of envy—call it Hollywood Green. It keeps everyone in a state of unnecessary mental turmoil. That day there had been a story in the paper about Suzanne Sommers, who believes that "her $30,000-a-week salary is below that of comparable stars on other successful television series." Invidious comparisons hide our blessings.

"Even so," Stein added, "almost all these people really do get cheated one way or another, no matter how much they are paid."

Ben Stein commented on the conspicuousness of my East Coast rig (call it faded Brooks Brothers), and recommended I pay a visit to Giorgio's on Rodeo Drive—if only to see how the rich dressed. Ben Stein barely touched his food, while I dutifully finished off a plateful that might have made an Orson Welles think twice. I told him I might have to do something about my waistline before visiting Giorgio's.

"Next time I'll tell you about how secretaries in Hollywood can become film producers within two years of arriving in town," he said. I begged him to tell me right away. He said that secretaries with distracted or absent bosses often find themselves running the place. They get to talk with stars on the telephone and the stars (suckers for fantasy) become intrigued by the voice on the line. At the same time, scripts are coming in for the boss to read—and he doesn't have time or isn't interested; or is worrying about a hundred other things. So the secretary tells the star that she has this script that would be perfect for him, really just perfect . . . (he has been complaining to her that he is always given these dumb parts that fail to do justice to his great acting genius) . . . and before you know it, she's in business as a producer!

Outside, the sun was beating down and someone had left the oven on in my car. The parking attendant had switched radio stations, in the half-minute he was at the wheel, from my dowdy all-news to FM rock. I drove to Rodeo Drive, a smart street in Beverly Hills, where someone was dusting off the sidewalk with a feather mop.

Giorgio's has a bar, and numerous photographs of the stars who have patronized the fashionable haberdashery. Among them was a framed photograph of Justice William O. Douglas and his fourth wife Kathleen. One had not thought of *him* as a Hollywood habitue. (Or had one?) I decided against buying a cashmere sports coat for $875. A man hunting through the racks next to me bought five shirts and a couple of sweaters for $980. He paid for them with $100 traveler's checks. When I mentioned this to someone later the only thing that surprised him was that the customer was American. Anyway, I could see why the Hollywood salaries are so high.

THE LAFFER CURE

A persistent myth about America is the belief that its wealth can be attributed to "abundant natural resources." In fact, there is no natural wealth—only elements. Los Angeles illustrates the point. It is reclaimed desert. It was desert until the 19th century, and it would almost certainly still be desert today if it were not part of the United States. Natural resources didn't build Los Angeles, mental resources did: energy and effort, pride and envy, faith and hope. The city is a triumph of mind over matter.

It is the small group of economists and journalists called supply-siders who tend to stress such views. Appropriately enough, the supply-side guru, Arthur B. Laffer, lives in Los Angeles. So I set off in search of him. He occupies a tiny office in the University of Southern California's Hoffman Hall. Laffer is 40, but he looks younger—a bouncy fellow with a windswept hairdo. He shared his office with numerous toy parrots and figurine turtles. On a nearly bare bookshelf behind him were *Cactaceous Plants* and *Turtles of the World*. There were one or two books about economics, but I have never met an economist who seemed to take himself less seriously, or was more removed from the milieu of dull professionalism that economists normally cultivate as though their reputations depended on it.

He is famous, of course, for the Laffer Curve. He said he knew of only one other economist with a curve named after him—A. W. Phillips of the London School of Economics. But when I changed the subject to dieting, Laffer soon lost interest in economics. He asked me if I wanted to know about the Laffer Cure. How Art Laffer Lost 50 Pounds Without Jogging?

I was all ears.

"Okay," he said, suddenly the brisk schoolmaster. "How do you maximize profits?"

"Ummmm, hold down costs?"

"And?"

"Uhhhh. . . ."

"Increase revenues. So I figured that if each pound is about 3500 calories, then what you have to do is maximize the difference between what you take in and what you use up. Principally, we use calories to maintain body temperature. Eskimos use up a lot of calories. Therefore I swam in Lake Michigan for two hours every day."

"And?"

"I lost 50 pounds in 40 days. Altogether I lost 75 pounds. My weight went from 235 pounds to 160."

Don't get hot—by running. Get cold. The Pacific Ocean will do just as well as Lake Michigan.

Next I asked him if he could explain the Laffer Curve.

"Okay," he said, bouncing up and down in his chair. "Robin Hood sets up a toll booth on the Trans-Forest Freeway, at the entrance to Sherwood Forest. You can drive round the forest but it's more convenient to use the freeway. Robin Hood collects a high toll from the rich and gives the money to the poor. The toll is equivalent to the tax rate. Okay? But when the toll is too high, what happens? The people in the Cadillacs avoid the toll booths. They go around the forest. Or they hire guards. So toll revenues begin to drop. Then Robin Hood lowers the toll. What happens? The Cadillacs come back, and what happens to the revenues?"

"They go up!" I said, playing teacher's pet.

I asked Laffer what the reaction of his professional colleagues had been to his sudden fame, or notoriety.

"I encounter substantial amounts of aggression from economists," he said. "I don't judge motives, but when you're with them you can feel it." One called him the "Laetrile of economics."

According to Jude Wanniski, the author of *The Way the World Works*,

Laffer first drew the much-loathed or much-admired curve on a paper napkin in the Two Continents Restaurant in Washington, D.C. I asked Laffer about this. Oddly, he doesn't remember the episode at all. Besides, he said, the Two Continents doesn't have paper napkins. It has linen ones, and he is quite sure he would never have drawn anything on a linen napkin. He was properly brought up, he said—always taught "never to draw on linen." He went to a prep school in Cleveland, then to Yale.

Laffer aroused in me the suspicion that it was Jude Wanniski who first drew the curve. Certainly he was the first to publish it. My theory is that Wanniski, a journalist, knew that economists' curves are not taken seriously unless they are drawn by people with Ph.D.'s. So he generously, or shrewdly, imputed it to his friend Laffer.

Eyeing the stuffed macaw perched somewhere above me (Laffer has half a dozen live ones at his home in Palos Verdes), I asked him how he felt about the theory of evolution—recently contested in the California school system.

"I have a hard time conceiving in my mind the development of the wing," he said. "So many accidental mutations would have to occur at the same time. A bird can't fly with half a wing." Laffer prefers the Lamarckian to the Darwinian hypothesis. If you *try* to achieve a certain result, then perhaps your offspring are more likely to inherit the right genes.

Incentives again.

LA BREA TAR PITS

Fred Evans, who is a black Jehovah's witness, 38 years old, knows the popular names of all the shrubs and trees in Hancock Park surrounding the La Brea Tar Pits. In many instances he can give you the Latin names, too. He was seated underneath a thuja tree (*arborvitae*), eating his lunch, and he was wearing the olive green uniform of the Los Angeles County Natural History Museum. He is the senior grounds keeper in the 20-acre park where he supervises ground crews doing park maintenance work, such as pruning, edging, planting and clearing. He is relaxed and friendly, finding it hard to imagine that anyone in Los Angeles has a nicer job. He attended a trade technical college, studying advanced turf management, and he also took extension courses at El Camino College.

From his position underneath the thuja tree, with the sun shining, the birds singing and the automobiles faintly humming on Wilshire Boulevard, he pointed out the Sequoia sempervirens, the Catalina cherry, oleander, eucalyptus and Washington palm trees—both *filifera* and *robusta*. It wasn't long before he had something to say about the evolution vs. creation controversy, which he had been hearing about in the news.

"Take a broom," he said. "No one would accept that something as simple as a broom just popped up by chance. You know that someone made it. But when you consider what plants and grass do with the energy coming in from the sun . . ." He paused and shook his head as he contemplated the mysteries of photosynthesis. "It's impossible to believe that it all just appeared by chance." By way of illustration he pointed to a nearby bank of Kikuyu grass, imported from Africa in the 1920s.

"I believe our young people should be taught that there is a creator we are responsible to," Fred Evans continued. "Once we recognize that there is someone we have to answer to, that will make us treat one another better. Otherwise you treat people the way you think best. We have a generation of young people today who feel that they do not have to answer to anyone. It has never been like this in the past—and I have to think that the theory of evolution is partly responsible."

He looked over at the George C. Page Museum, only a few yards away, where they show a film, slides and displays of the fossils trapped in the surrounding tar pits. It's not tar, incidentally, it's crude oil—decayed marine animal life, supposedly, but we're not even sure we know what oil is. "Of course," Evans said, "I don't know what they say in there . . ." his tone implied that if the Authorities disagreed with him that was their privilege but it wouldn't make any difference to him.

Twelve thousand years ago, mammoths, mastodons, short-faced bears and saber-toothed tigers roamed the environs of Wilshire Boulevard. Many of them were somehow trapped in the crude oil and never escaped. A hundred tons of old bones were recovered and some of them are on display in the museum. The oil is still seeping up, and can be seen in pools in low-lying ditches.

The interesting thing is that although the voice of scientific authority does strenuously disagree with the voice of Fred Evans, the straightforward facts perhaps do not. For example, the subsequent disappearance of the mammoths, saber-toothed tigers and others, not only from Wilshire Boulevard but from the face of the Earth, remains something of a mystery. "No one knows what caused this extinction of great mammals 12,000 years ago," the musuem's film commentary admits.

Most people know that the far more remote dinosaur extinction, 65 million years ago, is mysterious, but the embarrassing truth is that hardly anything is known about extinction, even though 99 percent of all species that ever existed are now extinct.

"Climatic change" is an unsatisfactory explanation, because (as another musuem display points out), almost all the other animals that happened to be at large on Wilshire Boulevard 12,000 years ago—frogs, toads, lizards, turtles and fish—are still quite contentedly with us today.

Fred Evans has an idea what happened, however, and he is only too happy to submit it for our consideration. "You see," he said, "I believe there was a flood, a great flood, that covered the Earth. And they couldn't get all the animals into the Ark. They were only able to take a sample of existing species with them."

That would explain why so many of the larger animals didn't make it.

Fred Evans finished his lunch and carefully stowed away the wrapping paper. It was time to return to park maintenance.

L.A. PUNK

My guide to the Punk scene was John Prizer, a young film producer not well known in Hollywood. We stood in line with about a hundred others outside the Starwood nightclub on Santa Monica Boulevard, waiting to hear two groups called "X" and Fear. For a moment I thought those waiting in line with us must have escaped from some nearby Marine boot camp. They had rough crew cuts and wore black vinyl jackets. One or two had chains and padlocks around their necks or displayed swastikas.

They turned and stared at us: conspicuous intruders. Prizer and I were dressed like good Saturday night straights with slacks and jackets, ironed shirts without tie, hair neither too long nor too short. I wondered whether the haywire Marine recruits were planning to drag us into an alley to find out if we were lawyers or something equally disreputable.

"They don't believe in the perfectibility of man," Prizer said in a low voice.

He believed that here were the cultural indicators of a new trend that would soon inevitably be felt across the land. Nothing less than the abandonment of the Enlightenment was at hand! Gone at last was the old era of peace, freedom and protest. Only a few years earlier, Prizer said, the rock scene was swamped with people who believed that if only a general permissiveness prevailed, the police handcuffed and the Pentagon abolished, we would all end up in a happy heap of hippiedom: man not only perfectible, but finally perfected.

"They read Hobbes, not Rousseau." the producer whispered, the better not to be overheard by these bold innovators, somehow at the cutting edge of Western civilization. Perhaps they might disapprove of our analysis and decide to beat us up for our pains.

We went into the club, which was colored ultraviolet, indigo-ink, and black. You needed a flashlight to see your money. The place was filled with shaven-headed punks. Many were from Orange County, Prizer said. Behind the bandstand, were large, rude, roughly lettered signs for our appraisal: FAG, GAY, HOMO, LEZBO. Signs of the times! And no doubt further evidence of the collapse of the Enlightenment.

The group called Fear came onto the stage. First of all they spat at the audience a few times. Then they commenced the first number. A great mass of ear-splitting decibels came crashing out into the room. Horrible! At the same time a great swirling tumult broke out among the spat-upon shaven-heads standing in front of the band. A fight, I thought. But on closer inspection it was a weird ritual of mutual elbowing and shoving that accompanied every . . . well, I was about to say "tune." But it was more like a great mass of noise, with no pretense at melody. The group made Elvis Presley seem decorous—suitable for a command performance before the Daughters of the American Revolution.

"For my generation there was the draft," Prizer shouted as loud as he could into my ear, on the off-chance that I might be able to hear a word or two. "This is their tribal thing."

The next number was announced as "Let's Have a War," and was greeted with a tremendous shout of approval. "YAY!!!"

"We want to go to El Salvador!" the band-leader bellowed into his microphone.

"YAY!!!"

The drummer started up what seemed to be an amplified pneumatic drill and the band set off again down its raucous, unmelodious path.

Not to believe in the perfectibility of man may well be excusable. A band that feels the same way about music is less so.

Outside, the neon signs and the palm trees stirring in the breeze came as a relief. "If that is what we have in store, give me the Enlightenment," I told Prizer. I asked him if he thought the Punks really wanted to start up an anti-progressive counter-revolution; or simply wanted to shock the bourgeoisie, which of course is quite a challenge today.

"For some of them it's ironic," he said. "But some of them are also telling us that if there aren't certain constraints, that's the way people will behave."

COPPOLA PLAYS GOD

Francis Ford Coppola, who is filming *One from the Heart* at Zoetrope Studios just off Santa Monica Boulevard, has himself a pretty authentic-looking Las Vegas right here in Hollywood. The Famous Pioneer Club Casino, Hotel Fremont, Lady Luck Casino, keno, craps, dollar slots, sporting events, all built to scale inside Soundstage No.3, and lit up with ten miles of flashing neon, border chase, front-nugget and slip-slant lights.

The crew is filming the Bluestown Dance dream sequence, an amazingly elaborate shot, which uses up the full 4 ½ minutes of film in Garrett Brown's Arriflex Steadicam. The sequence consists of Teri Garr and Raul Julia running and dancing in and around 250 extras and automobiles on this reconstructed Las Vegas strip, enacting their dream night on the town. As they dance their ballet of romance, Garrett Brown dances backward behind them with his seemingly featherweight 70-pound camera trained so smoothly on the couple that you would think it had invisible wheels. And dancing backward behind *him* is Vittorio Storaro, who everyone says is this tremendously famous cameraman, having won the Oscar for his work in *Apocalypse Now*.

But wait! Here are the actors, the extras, chorus girls, jazz band, Steadicam and cameraman doing this elaborate choreography, which will take all day to film—$100,000 is what this scene is costing in studio time and expenses alone—and where is the great man himself? Where is Francis Ford Coppola?

Perched on top of a ladder is Gene Kelly, whose old musicals have proved to be such a fertile field of study for continental cineastes. He is wearing dark glasses and a golfing cap, and he is pensively holding his chin between thumb and forefinger, studying the set like a golfer pondering a putt: a *direct link* between the musicals of the past "and today's musical generation," (says the Zoetrope press release). Actually,

Kelly's presence betrays a certain lack of self-confidence on Coppola's part. Despite his great success with *The Godfather*, Coppola seems to need this extra reassurance that he is qualified to make a musical. And so Gene Kelly was brought onto the payroll as a consultant.

Down below Kelly's step ladder are two canvas folding chairs, the indispensable props on any Hollywood set. One is for the cameraman and the other is for the director. But the former is occupied by Jean Luc Godard, the famous French director, another direct link with the cineastes and auteur-admirers whose good opinion Coppola seems to be seeking. Godard too looks pensive, even morose. When someone asks him if he is working for Coppola he makes an uncertain little teeter-totter with thumb and little finger.

Like God, Coppola is invisible. His chair is empty. He is off the set, in his silver-toned airstream Image & Sound Control trailer, exercising "total control over the environment's elements," (says the press release), monitoring the incoming videotape feed; planning, visualizing, even "previsualizing." In short, he is in total control of every detail, including the weather. This we are told is the new "electronic cinema," filmmaking's "greatest leap forward since the advent of sound in 1927," with direct links to the artistry of the past sitting right there before our eyes on the set.

Still, in some ways it all seems like a regression. Just about the time sound came in, Paramount sent Gloria Swanson off to Europe with her two dozen steamer trunks to make a film in the *exact locations* of the story. "A few years ago," a newsclipping related at the time, "no producer would have gone to all this trouble and expense: the picture was made right in the studio. . . . But the motion picture public is growing more discriminating every day. It demands authentic settings . . ."

No longer, apparently. The real, live Las Vegas is only a hundred miles away, but we're back inside the studio, albeit a totally controlled one. The principal actors and actresses, plus Vittorio Storaro and Garrett Brown are crowded around a small, black-and-white videotape monitor, minutely studying what may be the fifth replay of the laboriously filmed and refilmed sequence. This is supposedly one of the great technical breakthroughs—being able to watch on videotape what has just been filmed, without having to wait for the "rushes" next day.

It is getting late, and Teri Garr is tired. After watching the monitor at close range she lets her head fall on Raul's shoulder and says softly to him: "I hate it. I hate the whole thing."

Coppola must have heard her through some inconspicuous microphone. In an instant he is on the set—in person! This is reminiscent of *The Conversation*, Coppola's film about an electronics bugging expert who in turn is bugged and driven to insanity. Coppola surely overheard his leading lady's *sotto voce* vote of no confidence. He calls for a packing case, upon which to sit, and he sits Teri down beside him. He proceeds to play the old-style movie director, massaging her back as she sits there, and whispering into her ear. Soon he has her laughing. "Are you listening to what he's saying?" she says. " '. . . a little girl taking her first tap-dancing lesson . . . Gene Kelly . . .' " Here she turns to look for the famous old dancer, as though recognizing his presence for the first time. "I don't want it!" she says, reasserting her mood.

Coppola stands on a chair and addresses the assembled multitude. He calls for 30 seconds of total silence, like a prayer, before the final take of the day. Then he walks off the set. It is a dignified walk. He is a wee bit portly, but somehow stately in his straw hat and mauve shirt with loosened tie. As he walks out into the night he is followed closely by the Zoetrope Videotape squad, two young men carrying video equipment who have been making a film of the film since it all began last June.

Sometimes I think that *this* might be the film to watch, because if they splice together all the right segments they will surely show us the videotape relay centers, the computerized equipment, the airstream trailer, the ballet of the Steadicam, Jean Luc Godard and Gene Kelly—the behind-the-scenes story that somehow dominates *One from the Heart*, the story that is being filmed officially at Zoetrope Soundstage No. 3.

GHETTO MIRACLE

Technically, Mount Zion Missionary Baptist Church isn't in Watts, but it is in the area of south central Los Angeles that people think of as Watts, and it is where some of the Watts rioting occurred in 1965. It is here that the black pastor of the 2000-member congregation, Edward V. Hill, 47, climbs into his 1977 Lincoln Continental and drives past the squiggle-paint-sprayed, boarded-up stores, the long-junked sidewalk refrigerators and the ghetto-proof architecture (tiny barred windows eight feet above the ground), on his way to the World Christian Training Center on South Broadway and 47th Street.

But the Reverend Hill isn't your run-of-the-mill ghetto pastor. He is a member of the Moral Majority and a good friend of Jerry Falwell, Billy Graham, Nelson Bunker Hunt and Bill Bright of the Campus Crusade for Christ. He refers to Falwell as "Jerry" and a picture of Billy Graham hangs in his office.

E. V. Hill grew up in Seguin, Texas, during the Depression. He was poorer than ghetto kids are today, he says, and he feels that with due diligence and "discipleship in Christ" they too can prosper. He has been Mount Zion's pastor for 20 years.

"I felt and feel that there needs to be a loud voice calling for Biblical morality," Hill said in his office the other day, sipping occasionally from a tall glass of orange juice.

"I have several fears," Hill continued. "I earnestly believe that the U.S. as we have known it can fade away, and we have all of the signs: the destruction of the family, declining faith in government, ever-increasing government control over everything, and hostility between citizens and police. In my judgment, a democracy cannot long survive that way. It's a fragile system. In light of all that, how can anyone object to a strong call for morality? The fact that the call is now coming from men who have not heretofore been known as our friends makes no difference. If we continue to search out differences from the past where is the reconciliation?"

Falwell and others may have opposed civil rights and Dr. Martin Luther King Jr., in the 1960s, he went on, "but they've all made public retractions. And Jerry is now using the same methods as Dr. King, you see." So the Rev. E. V. Hill decided to join the crusade. He is a Republican, and he headed the Black Clergymen for Reagan Committee.

"I've never trusted white liberals," he said. "In my experience in the South they called us mister, but they left with the jobs and the power. And I find them to be much more vicious when you come up with an independent thought. They knew what was best for us! If you disagreed, they quickly called you an Uncle Tom. And they had the power to get all the other Negroes against you."

No longer, apparently. Hill's World Christian Training Center is organizing local church members into "soul winners" who will go forth and "win blocks for Christ through block workers."

Evangelism is the key, distinguishing Hill's from earlier wars on poverty. The poverty program degenerated, Hill believes, because "the people in Washington soon decided that they knew what Watts needed, funded it 100 percent, wrote the guidelines and ignored the local leadership. So we have no community way of monitoring what's going on. Does welfare break up families? If there was some local participation we might be able to find out."

The Rev. Hill plans to address the problem by "forming a union with 14 suburban churches" and investigating the matter.

The World Christian Training Center also runs a CETA-funded job training program for youths. Hill took me to meet the class, which was gathered in a back room. He told me that only a few days earlier some of them had been "gang members you would have been afraid to meet." They sat in disciplined, attentive silence as Hill addressed them briefly about attitude and motivation.

He told me later that they had a job placement rate exceptionally high by CETA standards. There's no mention of God in the program, though. The Rev. Hill wanted me to know that. Federal money is resolutely atheistic, and Hill's people have to worry about government plants enrolling in the program and informing on them in the event of religious indoctrination. But an aide in the hall told me that they do have "Christian instructors," and that, he said, is the secret of their success.

For confirmation, I asked the Rev. Hill how he accounted for the evident success of the program. He looked at me cautiously, perhaps assessing the possibility that I was a CETA spy from Washington.

"It's a miracle," he said.

SANTA MONICA BLUES

Diane Amann came to Los Angeles a couple of years ago from Libertyville, Illinois, planning to do graduate work in political theory at UCLA. But real life politics soon occupied her more than academic theories. First she worked in Carey Peck's unsuccessful campaign against Congressman Robert K. Dornan. Now she is press secretary for Santa Monica Renters Rights, an umbrella group which includes Tom Hayden's Campaign for Economic Democracy. Four Santa Monica City Council seats will be contested by the Renters Rights group on April 14th.

She was kind enough to drive me around Santa Monica, pointing out the sights. On the surface, Santa Monica looks like a pleasant enough seaside resort, but Diane gave me the impression that it was a problematical, indeed hazardous place to live, with concealed shoals just beneath the surface, and unexpected snares that made a mockery of the American Dream: lifeguards contracting cancer, for example; a shopping mall entirely out of keeping with "the small-town atmosphere"; water wells closed because of unnamed hazardous substances; rampant commercial development; crisis in the sewer system; a local airport about to be closed, and so on.

The redeeming feature of the place, I gathered from Diane, was rent control. Two years earlier the city had voted for a strict rent control measure and established a Rent Control Board; permitting the landlord only small rent increases (a percentage of the down payment for the building, which might have been quite small in the pre-inflation 1960s). Even so, the Renters Rights candidates vow to tighten up rent control even further.

I asked Diane if she lived in Santa Monica. No, she said, she had spent ages looking for an apartment but never could find one. "A friend of mine noticed a yellow sticker on an apartment door," she said, "meaning that someone had died. I went to the realtor to find out if it was available, but 12 people were ahead of me even though it had not been advertised."

According to Charles Isham, executive vice president of the L.A.-Western Cities Apartment Association, the number of apartments in Santa Monica has declined from 37,000 to 34,000 in the two years since the rent control ordinance passed.

Diane took me to meet the Reverend Jim Conn, one of the Renters Rights candidates, whose United Methodist Church in Ocean Park is a veritable hive of youth rapping, community interfacing, consciousness raising, liberation theology, and gay encounters. A brochure by the door proclaimed the joys of "Backpacking for Lesbians." A Sunday morning "experience," not service, was advertised.

"We have tried to redefine things for people," said the tightly blue-jeaned Conn, aged 36. "Traditional ways aren't terribly relevant and so we use a different style. To talk about worship. . . . Well, people don't find that very comfortable. The other word is 'celebration,' but that has become such a cliche."

A dance troupe was setting up to rehearse in the church. We appeared to be in the organ loft—minus organ.

The Rev. Conn told me that he favors social justice, economic justice and "liberation," but that he steers clear of the socialist label, because it has "too much baggage." He is strongly for rent control. "It gives people in a powerless position some sense of parity with their landlord," he said. "They still don't own but they aren't victimized."

But what about Diane not being able to find a place to live? How much parity does one have on the sidewalk? On our way back to the Renters Rights office, which the group rents from a member of the Santa Monica Rent Control Board, Diane pointed out the Sana Monica Shores apartment building. She said she believed the owner was refusing to rent newly vacated apartments. (True, as it turned out. The owner wants to test the theory that it is legal to go out of business in Santa Monica.)

Diane introduced me to Dennis Zane, 33, who worked for Hayden's Senate campaign and is a founding member of the Campaign for Economic Democracy. I asked him if rent control led to a decline in apartment contruction.

"You've got it the wrong way around," he said. "The decline in apartment construction preceded rent control. When apartments are no longer built you get rent control." He too is a Renters Rights City Council candidate.

"That's a familar cry from advocates of rent control," said Charles Isham, who is in the unpopular position of representing the interests of apartment owners, or landlords, in their bitter struggle with Santa Monica's City Hall. Isham himself owns two 12-unit apartment buildings, one in Santa Monica and one in Los Angeles. "Investors will run from apartment construction at the very mention of rent control," he said. "There was talk of rent control here several years before it was voted in."

The Santa Monica apartment association has 4,500 members. The average owner is a very small businessman with 12 units to rent. "Typically," Isham said, "owners are in their mid- to late-fifties, a good many of them black, Mexican-American or Oriental. They are people who sold their own homes after their children grew up and invested the proceeds in a small apartment building, which they manage and live in. They feel that they have been cheated by the changes in the law."

We don't hear much about these people because they are owners—capitalists. Also they are heavily outnumbered by their renters. For this reason the politicians have pandered to the tenants. In the short run (by the next election) the politicians win. But in the long run Los

Angeles will be badly hurt by rent control. (Los Angeles "stabilized" rents in 1978, allowing seven percent annual rent increases and vacancy decontrol.)

After an all-too-brief investigation, I have come to the conclusion that there are *no* privately financed apartments under construction in Los Angeles today (not just in Santa Monica). According to records in the L.A. Department of Building and Safety, permits for 443 buildings in the apartment category were issued in 1980. But a department official who wanted anonymity told me: "You'll be lucky if ten percent of the permits shown as apartments really are apartments. Possibly none are. They are condominiums, whose builders don't yet have approval from city planning for condo construction. Approval takes many months, even years. Meanwhile, they are issued apartment permits."

According to Isham, no new apartments were constructed in Los Angeles in 1979, and he believes none was built in 1980 either.

Rent control in Los Angeles has its roots in the no-growth attitudes of the 1970s. Deferring to this ideology, the L.A. City Council changed the construction rules, mandating low-density building where earlier high-density had been permitted.

This is how Charles Isham explained it to me: "I bought my 12-unit building in Santa Monica in 1977. If it burned down today I would not be allowed to rebuild it with more than six units, because now you are only allowed to build on 50 percent of the lot. Earlier it was 80 percent. Consider someone who has bought a parcel of land. The clock is running on his mortgage costs. He is only allowed to build six units. He also has to have underground parking and other features which make the building more expensive. To recover his costs, rents might have to be as high as $2,000 per month. But people can't afford that. But they will buy condos because of the tax break and the equity buildup. So the changes in the law created incentives for condos and disincentives for apartments."

Apartment construction fell off, continued demand (minus new supply) drove up rents, and rent control followed. Thereupon apartment construction halted completely.

Since money is mobile but buildings are not, developers are taking their money elsewhere and some owners have already abandoned their buildings, as in New York. By contrast, Houston today has an apartment glut. In the long run, such anti-property legislation will sap Los Angeles' dynamism.

But don't expect the City Council to worry about such distant events, because by then they will have been reelected and gone on to honorific retirement. And don't expect most of the people who already have houses here to worry either, because they rather like the idea of not having to share Los Angeles with so many newcomers. And that's the terrible thing about the no-growth, low-density zoning that is at the root of the problem. It pits those who have already arrived against newcomers. And guess which group has more political clout.

I went back to the newspaper and told a reporter what I had found out about low density and high rents. He listened patiently, removed a cigar from his top pocket and unwrapped it. "Tom," he said finally, with the good-natured air of one gladly passing on something he learned long ago about the human condition. "You know what the problem is?" Here he applied a match to his cigar and for a few seconds seemed to ponder the timeless foibles of man (which it was his duty to expose without fear or favor, calling the shots as he saw them). "You know why prices are so high?"

I looked at him expectantly.

"It's greed," he said, luxuriantly releasing a mouthful of cigar smoke. "Old fashioned greed."

ROBERTA WEINTRAUB: VALLEY GAL

Dan Van Meter is rail thin, sun-roasted, and weathered with age. He is wearing a bootstrap tie and metal tie clip given him by Jim Thorpe, the 1912 Olympic Decathlon winner, and he is sitting in the back seat of the chocolate brown Cadillac Seville as it drives down Woodman Avenue in the Van Nuys sunshine. He is worrying about the space shuttle, due to be launched from Cape Canaveral the next day. A few weeks ago, *Newsweek* magazine carelessly called Van Meter a "drifter" and ended up paying him $5,000 for the insult. Not that he needs the money.

As it happens, he was rummaging through a trash bin outside a liquor store in Van Nuys—looking for cartons and empty boxes for some reason—when he came across a woman's leg carefully wrapped up in a plastic garbage bag. He reported this to the police, who later discovered other parts of the woman—believed by detectives to have fallen behind in her cocaine payments to the Israeli Mafia. Three men were charged in the case and Van Meter will testify at their trial. Not that he knows anything about the Israeli Mafia, he hastens to add. All he knows about is finding the leg.

Far from being a drifter, Van Meter owns a couple of acres in Van Nuys "worth one and a half million bucks," he says. His uncle, Leon Douglas, was a partner of Thomas Alva Edison in the Victor Talking Machine Company. But right now his mind is on the space shuttle. If all goes well it will give us an advantage over the Russians. Therefore, something is likely to go wrong.

"I call them kremlins," says the old-timer.

"Instead of gremlins?" helpfully adds the driver of the Cadillac, Roberta Weintraub. She is the president of the Los Angeles Board of Education, and she has made quite a reputation for herself—heroine to some, villain to others—by opposing school busing for purposes of racial balance in the San Fernando Valley.

Dan Van Meter is one of her 250 volunteers, a small army of old reliables who show up at the office, man the phones and pass out leaflets in the anti-busing battle. On his lap he has a pile of Weintraub's "She Stopped the Buses" leaflets. Roberta Weintraub is up for reelection and she is driving to the Kaiser Hospital in Panorama City for an afternoon of campaigning. In her forties, she has orange-tinted curly hair, and is wearing blue and white spectator pumps with navy-blue stockings and skirt.

She says that she and her good friend and former Board of Education ally Bobbi Fiedler, now a Congresswoman in Washington, talk on the phone all the time.

"Bobbi says the Board of Education is a good preparation for Congress," says Roberta. "She says many of the other freshmen are hayseeds. They don't have the background on the issues she does."

"You'll be in Congress some day, gal," says Van Meter.

"No," she says, "I don't want to do that."

She stands in front of the hospital with an armful of brochures—a resolute, ambitious figure. "My husband probably thinks I'm nuts," she says, "because I don't have to do this." She has been called every name under the sun. But she says that her skin is "part alligator."

"Hi," she says brightly to hospital visitors, not fearing rejection. "I want to give you my brochure. The election is this Tuesday." Some walk a few steps and then do a double take. She always notices it when they do. And she is quite discriminating about strangers. Somehow, when the young liberals in faded blue jeans show up (there are not many of them), her back is turned.

"You can sense when people are against you," she says, adding that there is a certain type of liberal who wouldn't be on her side "even if I had sex with Mayor Bradley in the middle of Hollywood Boulevard."

"Oh," says a passing nurse. "You're the one I see on television all the time. I'll be! TV doesn't do you justice."

"You are the two thousandth person to say that," says Weintraub.

The nurse is a respiratory therapist today, but at one time she did facial makeup for the entertainment industry and so she has almost a professional understanding of the cosmetic effects of studio lighting.

She gives Weintraub a second look and says: "The problem is with the shadow on the bridge of your nose."

Weintraub expects to get at least 60 percent of the vote on Tuesday. "They elected me on a political philosophy," she adds, "not just busing." She and Bobbi Fiedler are "citizen politicians," she believes.

When she returns to her campaign headquarters in Van Nuys, the subject of her political ambitions comes up once again. Any chance of her trying out for Congress?

"It could happen, but I don't see it in the near future," she says. "Not if I want to stay married. What would I do with my husband? Take him to Washington?"

Tonight she is going out to meet the voters once more, this time to the bowling alleys. But right now she takes a rest, puts her feet up on the desk, closes her eyes, and gives an exploratory rub to the bridge of her nose.

CULTURAL MONUMENT NO. 15

From afar the Simon Rodia Towers in Watts are barely distinguishable from the medley of telephone poles and general urban clutter, especially in their present condition—under repair by the state architect's office and encased in untidy scaffolding. But when you stand beneath them in that sad, silent, abandoned neighborhood, with the disused "Red Car" railroad tracks running nearby, you realize what an extraordinary achievement they are. It seems hardly possible that one man built them. But Simon Rodia, an Italian immigrant who worked as a laborer for the city of Long Beach, did indeed build these hundred-foot high skeletal cones of spiral concrete and steel with "no helpers," no plans, no drawings, no formal education, and no money.

He built them over a period of 33 years, beginning after his 40th birthday, with his own surplus energy. Once they were finished, in the mid-1950s, he deeded the property, including his house and the towers, to a neighbor. Then he left the city and he never returned. And he made this wonderful remark: "I wanted to do something for the United States, because there are nice people in this country."

How remarkable, also, that these hundred-foot towers were built without the knowledge or permission of the City of Los Angeles! That tells you something about the expansive, unbureaucratized horizons of those earlier decades in southern California. In almost any other city in the world officialdom would have been knocking at Rodia's door as soon as the first upright was set in concrete.

Neighbors say that Rodia used to work, high up in the towers, in the middle of the night. They would hear him singing like a nightingale.

After he had gone, the city found out about the towers and immediately worried that they were a safety hazard. A municipal crane came rumbling down the street, and fortunately lost out in a tug of war with the towers.

It wasn't long before the forces of Culture and Enlightenment took over. Today the towers are designated Cultural Heritage Monument No. 15.

Instead of trying to pull the towers down, the officers of government are now trying to hold them up. The concrete—some of it 50 years old now—is beginning to crumble. And you hear rumors that the city may no more avail with its scaffolding than it did with its crane.

Nearby, meanwhile, stands the Watts Towers Arts Center, Department of Cultural Affairs, City of Los Angeles—a peculiarly inappropriate addendum to the legacy of Rodia. Here they have a museum "space," creative sharing experiences, rap sessions, community outreach, access, input, and "art activity sessions." Government-funded literature available in racks deplores the "recent decision to end the CETA VI job program," which will "deal a severe blow to women's arts."

Inside, I met John Outterbridge, the director of the center, who talked to me about diverse activities, community influences, potential problem areas and "resources"—the latter meaning money. Neatly attired in a three piece suit, he told me that he too was an artist—a painter and sculptor. And like Rodia he had worked for a municipal government—in his case Chicago. When he came to L.A., he told me, he was offered a job in a Compton poverty program; that in turn led to an offer of a salaried job "in a community situation." And here he was today, his card reading: Artist/Director, Cultural Affairs Dept.

We went out to look at the towers in the setting sun. As we traversed the short distance from the bureaucratic "space" to the scaffolded towers, I felt more strongly than ever that arts funds can do nothing whatsoever for creativity. On the contrary, they will almost certainly destroy it. As I looked up at the towers I was moved by Rodia's words: "I no have anybody help me out. I was a poor man. Had to do a little at a time. Nobody helped me."

Seeming to read my thoughts, Outterbridge gestured deprecatingly at his formal clothes. "Normally you would see me with dust on my feet from my sculpture and my work," he said. "At some point I would like to be with myself more. But I don't get to do that too much because of my responsibilities here—which are very important."

SCRIPT WRITERS ON STRIKE!

Not exactly a blue-collar strike, I said to myself as I stood outside 20th Century-Fox headquarters on Pico Boulevard. Members of the Writers Guild of America were patrolling up and down the sidewalk, just outside the gates. More like a blue-jean strike. A Calvin Klein denim and Lacoste tennis shirt strike.

As I stood there, an authentic Chevy-driving, hard-hat, blue-collar worker drove by, stuck his head out of the car and said to the effete strikers: "Go to work!" The calf-booted, corduroy-jacketed writers paid him no notice. They were hefting placards reading: "Studios Commit Unfair Labor Practices." Ron Austin, a guild board member and strike advisory committee member, turned to a fellow striker and said: "It'll be the same chaos we had in 1973," referring to the most recent strike of guild writers. "There are no Indians, only chiefs."

Some of these chiefs-on-strike are paid $200,000 for a film script. And that's not tops either. I heard half a million mentioned while I was out on the picket line. A guild member who was present when the ballot was taken on Friday night told me that the principal writer on

"General Hospital," a daytime soap opera, announced at the meeting that she was being paid $20,000 a week.

Whenever you hear about people going out on strike, you always have to consider the possibility that they are striking not because they are being paid less than the market can bear, but because they are being paid more. And they are looking to keep it that way—by keeping out rivals who will work for less.

That really is the function of a union, after all. It is a collaboration ostensibly directed against management, but in reality against workers: sometimes called scabs.

When you have a market with a small number of producers, it may be possible for those producers to band together monopolistically and pay the workers a subsistence wage. That is supposedly what happened at the time of the Industrial Revolution, giving rise to Karl Marx and his many allies. But a small number of producers also makes it comparatively easy for the workers to "unite," creating their own monopoly and driving up wages by the simple expedient of threatening violence upon any of their mates willing to work for less. And notice that unions, unlike management, are exempt from anti-trust laws. Thus does contract give way to extortion.

This works fine until there is a technological shift, or a regulatory relaxation, expanding the number of producers. At that point it becomes a complex and time-consuming business for unions to "police" new entrants into the workplace—to "organize," as it is so politely put.

In the entertainment industry, this is now happening with the advent of cable television. Cable and video will diversify the market, giving us many producers where before we had few—the TV networks, the movie "majors."

One reason why magazine writers such as myself have no union is that there are so many magazines. As the anti-trust experts say, there are no "barriers to entry" in the magazine business. Indeed, the legal phrase "barriers to entry" takes on concrete meaning when you see $200,000 script writers patrolling the gates of 20th Century-Fox. (Not that they are actually barring entry to the lot, but they'd like to.)

With all these new video and cable production companies appearing, some of the writers' guild people are afraid of discovering what their true worth is. Cable producers don't have contracts that exclude non-union writers, as they *are* excluded from the "majors."

And apparently some of the majors, such as 20th Century-Fox, are beginning to set up deals with the cable companies.

"We are fairly certain that there are already non-guild writers working for pay TV," I was told by Allan Mannings, who is on the board of directors of the guild.

"So what would happen if I wrote something for a cable company without being a member of your union?" I asked Ron Austin.

"We wouldn't sign a contract with them," he said.

"It seems to me you're organizing against other writers, not against management," I said.

"You're talking about all unions there," he said, "not just the Writers Guild."

"That's right," I said.

"Exactly."

Exactly. In the long run it's not people like the 20th Century-Fox management the guild is organizing against. It's people like me.

THE CLATTERING INFERNO

Now that I have the ear of Hollywood I'd like to register a mild complaint. I've been to quite a few movies lately—*Tess, Excalibur, Elephant Man, Raging Bull* and *Coal Miner's Daughter*—and although I usually enjoy what I see, I rarely enjoy what I hear.

And the reason is that I really can't hear much of the time. The words, I mean. I can hear the clattering sound effects. But the dialogue in today's movies is rapidly becoming inaudible. No one has ever accused me of being hard of hearing. I never have a comparable problem with the radio. But something seems to be seriously wrong with the soundtrack of a good many films. Some people agree with me heartily when I raise this issue. Others give me a blank look.

I challenge members of the latter group to repeat verbatim some of the dialogue in *Raging Bull*. Ditto a good many of Natassia Kinski's mumbled lines in *Tess*. (*Coal Miner's Daughter* was an audible exception.)

What's gone wrong with the talkies? According to Don Rogers, the technical director of the highly esteemed Goldwyn Sound Facility, now an adjunct to Warner Hollywood Studios, the problem is out there in movie-theater land. (He admits there's a problem, at least.) The theaters, he says, often have sound reproduction that does not compare with the recording equipment. This is like pouring high-octane space shuttle fuel into a Model T. You end up with a lot of sputtering, crackling and banging.

"Raging Bull" was recorded by Goldwyn Sound, and Rogers agrees there was a problem with its soundtrack. "There was very little revoicing, or looping," he said, referring to the lip-synchronized, retaping of dialogue in the studio after the film is shot. "Martin Scorsese wanted it that way, because he wanted everything real. Authentic. With maybe mumbling, for example. The way you would have heard it on the street. I agree it might not have been the perfect answer."

According to Ioan Allen, the vice president of marketing for Dolby Laboratories in San Francsico, another problem is the excessive use of wireless "lavalier" microphones. These have no wires and are worn concealed in actors' clothes. "They make filmmaking easier but the sound worse," Allen said. "Directors like them because they simplify life on the set. You don't have to worry about microphone booms and soundmen everywhere. But now the directors are beginning to realize the sound isn't so good, and they are backing away from them, thank goodness."

A microphone has to "see" the actor's lips to achieve a natural sound, Allen said. When it doesn't, because of concealment in clothing, high frequencies are lost and then artificially restored later by turning up the treble control. "The result is an artificial sibilance," he said, "an unnatural sound." He said the main problem with *Raging Bull* was an excessive dependence on lavalier mikes.

Another thing: There are too many of these confounded microphones—sometimes eight, according to Rogers—and very often they are too close to the sound source. The result is all that maddening crackling and amplified rustling you have to sit through nowadays. Some of today's sound engineers are too dumb to tune out these sonic pops and squeaks. They seem to regard all sounds as equally worthy of our attention.

More to the point, today's sound man is likely to be not so much dumb as deaf, having graduated to movies from the unwholesome school of hard rock. That's why the volume level is set so high in many of today's movies. Quite apart from perforated eardrums, an incurable professional deformation among hard-rock engineers dictates that sound equipment must show off its performance capabilities by having the volume turned up close to the maximum level at all times. For the rock-deafened, "good" means "good and loud."

The result is today's typical clattering inferno of sound effects, with the dialogue merely one among many contributors to the decibel level.

I'm surprised the Writer's Guild doesn't complain. Films these days more and more resemble "visuals" (often excellent), plus sound effects. It was striking that, despite its hard-to-hear dialogue, the plot of "Raging Bull" was perfectly intelligible. If that means films don't need words, Hollywood soon won't need writers.

People are beginning to complain, Rogers told me. More and more people are asking theater managers for their money back.

"Thank goodness they are," Allen said when I told him this. "Let me assure you that it is our intention to get the sound better than it is. But still the best weapon is to get people in the moviegoing audience to complain to theater managers. Unfortunately, many of them are still too comatose to do anything. But when they do complain, it does bring results."

L.A. FREEWAY

David Brodsly, who works for the city administrative officer, has spent the last three years, on and off, studying and writing about the Los Angeles freeway system. The result of his labor, L.A. *Freeway*, will soon be published by the University of California Press. Quite an achievement for someone who is still only 26.

As far as Brodsly knows, there are no true freeway buffs in Los Angeles, with the possible exception of Joan Didion. She has contributed a word of praise for the book's dust jacket. Whatever their area of interest, buffs usually only emerge after the topic at issue (steam locomotives, the Civil War, varieties of barbed wire) has declined or disappeared.

If so, there should soon be a lot of freeway buffs, because the Age of Freeways is now coming to an end. "They are not building any more freeways here," said Brodsly, as he drove me along the Santa Monica Freeway—the archetypal L.A. freeway. "The proposed Century Freeway was the last bet for a new freeway, and it looks now as though they won't build it."

The first local freeway, following a little behind New York, was the Pasadena Freeway, which opened in 1940. In the next ten years only fifteen more miles of freeway were built. Los Angeles, Ventura and Orange counties now have 715 miles of operating freeway. Almost all of those remaining 700 miles were built between 1950 and 1975.

In any complex system—whether nation, city or individual—deterioration sets in once growth stops. Only in the minds of planners does stasis exist. It will surely be the same with the freeways. Now that growth has stopped, decline will follow.

Remarkably, the state could afford to build new freeways until a few years ago. Now there may not be enough money to keep the existing system repaired. "I would say the freeways are undermaintained right now," Brodsly said. In 1963, the state collected seven cents and the federal government four cents on every gallon. These figures have not been increased since. Corrected for inflation, they are equivalent to two cents today. And gasoline use is also declining, thanks to more efficient cars. Nevertheless, Brodsly pointed out, Caltrans persists in its rhyming signs that urge even less gasoline use. This will result in even less revenue for the department.

Brodsly dispels a couple of freeway myths. They didn't cause the urban sprawl. Nor were they responsible for the earlier downtown decay. Both were well advanced before the 1940s. The sprawl was already here at the time of the trolley cars. The recent downtown rejuvenation was in turn made possible by the freeways, which make it easier to drive in from the suburbs.

One of the most interesting features of the freeway system is that it was only "planned" in the most haphazard way—and that mostly after the fact. The L.A. County "Master Plan of Highways," published in 1941, "did not contain plans for or explicit discussion of the freeway system," Brodsly writes in the historical section of his book.

That shows just how hard it is to see the future—even (as in this instance) after it had already begun. And if planners *had* foreseen the full system, "they would never have been able to pass the legislation needed to pay for it, because it would have seemed too expensive," Brodsly told me. "The freeway system emerged because it never got bogged down in master-planning." By contrast rapid transport—both the oldest and latest transportation panacea—has always been planned to death.

It's fashionable to deplore freeways. Leftists tend to dislike them because they "isolate" people in their private automobiles, postpone the collective Utopia, and grant intolerable freedom and mobility to the bourgeoisie—for whom public transport would be more appropriate. When he started studying freeways for his thesis at the University of California at Santa Cruz, Brodsly said, his professors took it for granted he would adopt the customary negative stance. To his credit, he didn't.

In fact, I rather gathered as he drove me around the freeway system that he has a sneaking admiration for this mode of transportation. Don't we all?

STICKY WICKET FOR HOLLYWOOD CRICKET

I made it to the Sir C. Aubrey Smith cricket field in Griffith Park, home of the Hollywood Cricket Club, just in time. After exactly 50 years of cricket, nestled in the Hollywood Hills, this was the last day of play. They say that, after the club's first meeting, P.G. Wodehouse wrote up the minutes. Boris Karloff played here. So did David Niven, Errol Flynn, Basil Rathbone and Nigel Bruce. So did Don Bradman, the Babe Ruth of cricket. He came here with the touring Australian team in 1932.

Now they are about to turn the field into an equestrian center. Unspeakable dressage will replace the hallowed ring of bat on ball. The Los Angeles Department of Parks and Recreation has leased the 70 acres of land, just north of the Ventura Freeway, to Equestrian Centers of America. The bulldozers are scheduled to arrive this week. Perhaps they are already slicing into the well-watered, mown and rolled turf.

It was a sad moment as the Hollywood team (fielding) and the Orange County team (batting) came in for tea. The seats and benches near the old pavilion had already been taken away. In the old days, well dressed spectators in deck chairs would ring the boundary. Gladys Cooper and Olivia de Havilland would serve tea. Hollywood was still doing its British imitation in those days.

Now the cricketers are mostly from British Commonwealth countries. Not many white faces, if the truth be told. And even fewer Englishmen.

Cricket is a growing sport in Los Angeles, though. There are 16 teams in the Southern California Cricket Association—800 cricketers in all, according to Alan Rowland, vice president of the association. Rowland told me that an earlier attempt to replace cricket with polo was beaten back with a canny call to City Hall: "How many blacks play polo?" the pol was asked. No more was heard of that scheme. Alas, the current cricketers, although they include many blacks, seem to lack political clout. Or the equestrians have an excess of it.

Some disconsolate Pakistanis came out of the pavilion carrying styrofoam teacups.

"It's a good thing Doc didn't live to see this," said Leo Magnus, who came here from Jamaica 20 years ago. Doc Severn was a Hollywood Cricket Club stalwart, who died a few months ago, aged 90. (He played into his 80s.)

"Sad day today," said Ian Wright, a Yorkshireman and vice president of the club. "Seems impossible, doesn't it? A wonderful bit of greenery in the middle of Los Angeles." Wright, a photographer for the London *Sunday Times*, soon to publish an article on Hollywood cricket, has collected many old photographs of the field in its heyday. Not only was this the 50th anniversary of the club but the 100th anniversary of cricket in Los Angeles.

"An oasis," agreed Magnus.

"That's a good way of putting it," said Wright.

"A whole lot of work we've done on this field," sighed Magnus.

There is a small handful of American cricketers, one being Derek Rowland, Alan's son, who was born and raised in Van Nuys. He made an excellent catch in the outfield. (Barehanded, baseballers please note. I've been meaning to tell Bowie Kuhn for some time now that baseball could be improved by taking the gloves off the outfielders. Derek Rowland agrees.)

"I've heard rumors that the money didn't come through," said Magnus hopefully, referring to the equestrian deal. It's been on-again, off-again for years, apparently.

"The story may be exaggerated," agreed Bob Beeney, who batted for Sherborne School in the 1950s and is now in L.A. real estate.

But I called James Hadaway at the Parks and Recreation Department and he said: "I understand they're starting to break ground today."

Sheldon Jensen of Administrative Services said: "They'll be in this week. I think they're going to fence it off at the end of the week."

The Hollywood Cricket Club won't go out of existence. They will continue playing at the Woodley Cricket Field, near the intersection of the Ventura and San Diego Freeways. But the games will have to be moved to Saturdays, because the Woodley field is already booked on Sundays as a result of the expansion of the league. So does anyone have a spare field, preferably flat and centrally located? It will be mown, watered, rolled, and brought up to English country house, front-lawn standards. Well, perhaps not quite.

Oh, the result. Orange County won, by nine wickets. Kim Foster of Orange County, originally from Sydney, Australia, scored 41 runs and was the hero of the last hour.

San Francisco,
Democrats

On the day before the convention I ran into Lili Osborne, Eric Hoffer's long-time friend, a block or two from the Moscone Convention Center. She was waiting for the AFL-CIO protest parade, which was due to come down Market Street any minute. Hoffer died over a year ago, in the spring of 1983. Lili told me that she had joined the Seafarer's Union when she was a tuna- and sardine-packer in Monterey. Her former husband, Selden, had joined the longshoremen in the 1940s, which was how he and then Lili became acquainted with Eric Hoffer.

"It's sad that union's are no longer representative of what they set out to be," Lili said, gazing out at the marchers. "Maybe they've had their day, I don't know." She didn't elaborate, but you couldn't help noticing the Adidas-shod, pale pink preppie-shirted yuppies in among the blue collars and railroad hats.

"No U.S. Intervention in Central America!" read the typical college-grad interloper's placard. This seemed odd in itself, since trade unions and military spending are clearly symbiotic. But then yuppie Democrats do not really have trade union interests at heart. As Michael Kinsley wrote a few weeks ago in *The New Republic,* supposedly upwardly mobile yuppies are frequently headed in the opposite direction, finding it increasingly difficult to live in the style that their parents took for granted. They are not self-consciously joining the working class, like Selden Osborne when he graduated from Stanford, but tumbling into it willy-nilly; marching along with it, protesting Reaganism with it at the outset of the Democratic National Convention.

Next came the gay/lesbian protest parade, coming down Market Street in the opposite direction. There was something really depraved about this collective act of defiance—an attempt to represent choice as victimization ("We demand massive federal funding to end the AIDS epidemic!") and to neutralize opposition by selective appeal to Christian virtue ("Bigotry Is Immoral!")

San Franciscans took it meekly on the chin as the 50,000 to 75,000-strong phalanx came cawing and crowing and shrieking down the street. Many bystanders were no doubt affronted but were too outnumbered and too afraid to register any protest. The conspicuous exceptions were four or five young men from Long Beach, who stood on a corner carrying tremendous, sail-sized banners, for example: "No! No! Homo!" and "Read The Bible While You're Able!"

Their leader, a man in his thirties with a cropped beard, wore a crash helmet. He said his name was Bob Bible. Around him was a tremendous, indescribable din of catcalls and whistle-blowing as the enraged marchers came abreast of Bible's provocative signs and responded with hoots and howls. Bible lifted his bullhorn to his lips and his admonition, "Read the Bible people," came reverberating loud and clear across Market Street. "You haven't been reading the Bible."

In response there was a steady, rhythmic chant: "Gay rights now, gay rights now!" A group in the parade carried a street-wide sign: "Enola Gay Faggot Affinity Group. Gomorrah for a Tomorrah."

"That's the good thing about America," said Bob Bible, his amplified voice echoing off the surrounding buildings. "Your rights are protected. God bless America! Show God you appreciate it by reading the Bible." He didn't look like a Bible-Belter at all: he, and his companions, might have been hippies who had tired of the aimlessness of hippiedom, or could even have belonged to a motor-bike gang. They were certainly fearless.

By now Bob Bible was surrounded by five very large policemen in motorcycle gear.

"Thing is, we've got 50,000 people here," one said very close to Bob. They were standing crash helmet to crash helmet and speaking in conversational tones. "We've got this problem with people who might get violent."

"I need to preach, don't I?" Bob said.

"Wouldn't it be a good thing to have your own parade. . . ." As the cop spoke (and there are no name tags on these San Francisco policemen) Bob put his bullhorn up once again and called out, "Repent!"

The cop took the bullhorn and examined it carefully, as though looking for design flaws, or as though it might contain a hand grenade. He eventually found it to be unconstitutional. Bob Bible lacked the necessary permit.

"Can you do this some other time?" asked the policeman. "Could you be Christian and do what we ask you to do?"

"The Bible says they're an abomination," said Bob. And again he called out "repent," but this time, unamplified, no one could possibly hear him.

"Why don't you forget about it?" said Badge No. 1674. "Can you do this in the name of Jesus?"

And that was the last we heard from Bob Bible. Toward the end of the parade came excommunicated Mormon and presidential candidate Sonia Johnson, and a small cluster of the feminist faithful. And behind her, behind the AIDS patients dressed in white, was a motley assortment of clergymen from various denominations.

That evening I went to a smart media party given by *The National Journal* at the Circle Gallery, close to Union Square. Almost the first person I saw was Michael Kinsley of *The New Republic* and author of its TRB column. He was wearing a cream-colored suit and looked more like a lawyer than a journalist, most of whom were wearing regulation woodwards—navy blue blazer, grey slacks, and button-down shirt with tie firmly in place (top button undone for investigative work only). I asked Kinsley if he could think of a media angle to the convention. Everyone was saying that there were 15,000 media people here for the convention, outnumbering the delegates four to one. And everyone was saying that this was a deeply depressing state of affairs, but they didn't really seem to be depressed when they said it.

Kinsley pointed out that whereas the media were being paid to attend the convention, the delegates had come at their own expense.

Somehow this surprised me. I knew I would never have thought of it myself. Later on it occurred to me that it belied Elizabeth Drew's complaint, frequently voiced in *The New Yorker*, that too much money is flowing into politics. If the Democratic Party can't afford to pay for the hotel rooms of its delegates, it would seem that there's not enough money. (In fact, some delegates—those swearing fealty to special interest groups such as Big Labor— were financed.)

We were joined by Robert M. Kaus, formerly of the Hollings campaign, and Jonathan Alter of *Newsweek*, which made us a *Washington Monthly* alumni quartet. Then along came Michael Barone, co-author of

The Almanac of American Politics and a *Washington Post* writer. He shared some of his findings with us. My goodness, Geraldine Ferraro had only just been named vice presidential candidate by Walter Mondale, but Barone had already been over her Queen's district in his seven-league walking shoes, up and down Queen's Boulevard, up to her office, back and forth, constituents interviewed, a thousand bits of data collected and by now probably stored away safely in a computer in The Washington Post building.

Here we were still lazily in our starting blocks and Barone had already completed the first lap. He asked me what I thought of the Ferraro choice. What I *meant* to say was that none of my conservative friends could understand what all the hullaballoo was about; that she was a pro-abortion left-winger who in a former age might very easily have been excommunicated from the Catholic Church . . . but what I stumbled out with was: "She doesn't balance the ticket. . . ."

"Interesting," said Barone, clicking the verdict into his computer. Then he strode off in search of more data. It seemed likely that within another hour or so he would be another lap ahead.

The next day Ferraro and Mondale came to an open-air welcome rally on Market Street. The crowd waited patiently for nearly an hour in sunny, breezy weather lifted from a French Impressionist painting. Finally the VIPs arrived, with much preliminary scouting and walkie-talkie communication by Secret Service agents. Geraldine Ferraro, thin-lipped and pointy-jawed, conveyed a sense of ambition bordering on the ruthless, I thought. There was a glint rather than a gleam in her eye. Beside her on the platform were San Francisco's leftist congresswomen, the unprepossessing Sala Burton, and the more trendy-looking Barbara Boxer, who also represents Marin County, to the north.

Ferraro read a dull little speech, through granny glasses. At the end of it she raised her fist, not her open hand. (But at this year's convention, unlike 1980, the rest of the Democrats waved open-handedly from the platform rather than with clenched fist. I take this to be a sign of health within the body politic.)

Joan Mondale (Joan of Art) seemed to have been dieting to the point of emaciation, poor thing. Her daughter Eleanor looked as though she was planning to attend some punk-rock event, with her loose pink tie and mid-calf grey pants. Then old Mondale himself came to the mike and said we should "get these god-awful nook-alar weapons under control." One had not realized that they were out of control. Mondale could have an effective campaign issue here if he can prove the point to everyone's satisfaction. Then he said we should "get these deficits down," hardly a stirring message.

Mondale seemed to be a nice enough fellow, though. Everyone says that he is "witty in private," and I could see that he had a sense of humor. I didn't get the impression that he was an ideologue. Mondale struck one above all as a deferential soul, perhaps having been rebuked once too often in childhood. He has an older brother who is a Unitarian minister— I met him purely by accident at a Washington hotel ballroom on the day of Jimmy Carter's inauguration.

Mondale looked like a man more comfortable taking orders than giving them—just like George Bush. This would be tolerable in almost anyone except the President of the United States. And of course one can have no confidence at all in the people who will be advising Mondale.

Again, when Mondale gave his acceptance speech on Thursday night you sensed in him a meekness that spilled over into sheepishness. When he delivered a line he almost seemed to be looking back over his shoulder in the direction of offstage aides, checking that he had read it as he was supposed to, that he was doing okay, that he still had everyone's approval back there, and that he really was doing the right thing for his country and his party.

When I went into the Moscone Convention Center later that day my heart sank: it was nothing but a great big milling scrum of about ten thousand people circulating about endlessly, each with a necklace of credentials, and most of them further festooned with badges and buttons. I caught sight of Morton Kondracke of *The New Republic* charging along with another journalist at his heels, Kondracke crying out to anyone who would listen: "Where's the Cuomo press conference!" Eventually a Cuomo aide appeared. He was swiftly engulfed in a globular cluster of reporters and cameramen and swaying boom microphones.

I was grateful I didn't work for the daily press. Imagine having to phone one's editor in the midst of this melee and being told that one had just missed some vital development—a Cuomo press conference or what- ever—that a wire service reporter had chosen to treat as "news." Hun- dreds, or thousands of reporters covering the same event tends to result in a uniformity of interpretation for this reason.

And perhaps that is why, as I soon began to realize, the presence of so many reporters really did have a dispiriting effect. Journalists were not only interviewing journalists, they were joking about it, apologizing for it, and deploring it. The periodical press, I found, were bivouacked several hundred feet from the podium. The *Newsweek* people had binoculars on their desks, and I was behind them. When I held my arm out straight, the tip of my little finger obscured the podium. So distant were we that the sound and picture from the big TV monitors were confusingly out of synch—like watching a man chopping a log across a valley and hearing the sound on the upstroke. Mel Elfin of *Newsweek* told me that at no earlier convention had he been so far from the action. Someone told me that the distance was 450 feet.

We were allowed onto the Convention floor for 30 minutes at a time. It was packed with delegates and gossip hunters, otherwise known as journalists. I bumped into Walter Shapiro, an old friend from *Washington Monthly* days, now with *Newsweek*. He was talking to Murray Kempton of *Newsday*. Shapiro made the most succint comment on the news media-convention relationship that I heard all week. "First the networks made the convention dull," he said, "then they abandoned them." He was referring to the much reduced network coverage this year. (Even so, I was told that CBS and NBC each had 600 people on hand, and each would be spending about $7 million on the event.)

I asked Shapiro to elaborate. Once upon a time, he said, the political conventions were uproarious and interesting and unpredictable. Network television then broadcast this agreeable pandemonium coast to coast, and everyone liked it except for the party leadership. So they took steps to restore predictability and order to the proceedings, which in turn became so dull that not many people wanted to watch them any more. And this was where we were today. Moreover, C-Span and the Cable News Network were taking over the gavel-to-gavel coverage, and as more and more people gain access to these alternative channels, the politicial junkies who actually enjoy watching politicians giving speeches will switch to these channels, thereby diminishing the networks' role even further.

It was perhaps significant that inside the press lounge, three out of the four TV monitors (sometimes all four) were switched to C-Span even during the periods of network coverage. The press doesn't need to be told by Dan Rather that Geraldine is an Italian ethnic from Queens, and I'm sure the same is true of political junkies across the land. C-Span thus becomes the first true "medium," like a clear pane of glass between the event and the viewer.

* * *

Frankly, the main floor of the convention center wasn't a very exciting place to be. I dutifully interviewed two or three delegates and scribbled down what they said, but I haven't even bothered to look at my notes because I know it would be too boring to repeat. It is hard to say anything interesting about an event that has already been covered live on television.

The great unexpected pleasure of the Convention Center turned out to be the Railroad Press Lounge, a huge, never-ending press freebie sponsored by the Association of American Railroads. This lounge, frequented by as many as 3,000 journalists a day, provided us scribblers with free food (ham rolls, roast beef and horse radish, pizza rolls, hot dogs), coffee, tea, Pepsi, Coke, and even an endless supply of free beer. Add to that comfortable chairs, TV monitors, and a couple of Clairol Air Massage "foot fixers," and you can well imagine that it was standing room only in no time.

What about Post-Watergate Morality and all that? Weren't the press falling into the error they were always imputing to others, namely accepting handouts, thereby becoming tools of the special interests, in this case Big Labor? In *The National Journal*, special daily editions of which were published during the convention, I read the following comment: "In normal circumstances the press might flinch from accepting unlimited amounts of free food and drink from a lobbying organization. But for hardship assignments like political conventions, the top priority is survival and that is why the lounge was so crowded on opening day."

Hardship assignment—that was it. I settled down in front of the TV monitor to watch the Mario Cuomo speech with a large cup of foaming Pilsner Lite, reassured that if Bob Woodward or some other member of the ever-vigilant *Washington Post* SWAT team were to come by taking names, I would be able to plead hardship. I was further reassured by Hendrik Hertzberg, the editor of *The New Republic*, and by Richard Cohen, columnist for *The Washington Post*, that this was an "old tradition" and was therefore immune, by tacit agreement, from latter-day media moralizing. And so I contrived to enjoy my beer and pizza, as I noticed a good many other members of the press corps did. Cohen added that, although it was true that railroads generally do get favorable press treatment, this was more a matter of "nostalgia" than of sold souls in the news media. I thoroughly agreed with him on this point and promptly refilled my cup with Pilsner Lite. If the Railroad Lounge was good enough for Richard Cohen, for Victor Navasky of the *Nation*, and for young Ron Reagan, covering the event for *Playboy,* it was good enough for me.

As for the substance of Democratic Party policy, what is there to say that hasn't already been aired and mulled over endlessly? Alex Brummer of *The Guardian* (English version, once based in Manchester) said what no American journalist today would dare say when he described Governor Cuomo as "looking like a Chicago gangster. . . . He looked like he'd just walked out of an Al Pacino film." The enthusiasm for Cuomo at the convention seemed to be uncritical to say the least.

When Cuomo waxed indignant about the budget deficit, I noticed, the hacks in the press lounge didn't even bother to look up from their food. They had tried for three years to stick President Reagan with this complaint, insincerely appropriated from the arsenal of fiscal conservatism, to no avail. The American people wouldn't listen to this siren song when the media sang it, and no doubt it wouldn't work for Cuomo either. The Democrats' new-found enthusiasm for reducing the deficit struck me as politically inept. The Republican Party got nowhere with it for 20 years at least.

The Democrats this year evidently felt obliged to disguise their increasingly leftist (called liberal) ideology with conservative labels and symbols: talk of "family" and even waving American flags—something they wouldn't have dreamed of doing in 1976. But it was obviously contrived and a lot of the delegates must have felt awful silly waving the Stars and Stripes. (The things one has to do to get ahead in politics these days!)

Governor Cuomo must have been delighted to find that he could use the word "family" to describe the entire country, and yet encounter almost no opposition or criticism from those who specialize in using words for a living—the press corps. You would think they might be more vigilant about the corruption of language. But they just don't notice such nuances, or care if you point them out.

"We believe in a single fundamental idea . . ." Cuomo bellowed out. "The idea of family. Mutuality. The sharing of benefits and burdens for the good of all. . . . We must be the family of America, we recognize that at the heart of the matter we are bound to one another, that the problems of a retired school teacher in Duluth are our problems. That the future of the child in Buffalo is our future," and so on.

The New York Times responded that Cuomo had demonstrated "that it is possible to carry the battle to Ronald Reagan on terms and territory that he had previously made his own: the terms of family and traditional values. . . ." Yes, it is possible, but only by subverting the meaning of words—with a little help from the "adversary press" that mysteriously transforms itself into the bodyguard press when the liberal agenda is at stake. "As an Italian Catholic," added *Wall Street Journal* reporters Dennis Farney and David Rogers, "he doesn't just talk about 'family' and 'flags,' he also seeks to wrestle them back as issues for Democrats." But what Cuomo means by "family" bears about as much relation to the traditional family as it does to the Mafia family.

Richard Cohen was a little bit more candid when he wrote in *The Washington Post* that what many people need "are political leaders who understand that for many the conventional family is a bygone thing. And they need policies to suit." Well, don't worry, Richard, the Democrats will provide them. "Family" was a codeword intended to confuse the majority who will no doubt imagine that Cuomo had the traditional family in mind.

Likewise, Geraldine Ferraro appealed to traditional sentiment when she said that "Americans want to live by the same set of rules." But she, and the Democrats, have no intention of instituting or supporting such rules. How can they, when they seek to encourage some Americans to live at the expense of others? How can they when they support affirmative action—an explicit recognition of the claim to privileged treatment by virtue of group membership? How can they when they seek to restore the progressivity of the income tax?

While on the subject of unequal rules intended to produce equality of outcome, it is worth repeating the interesting item reported by Al Hunt in *The Wall Street Journal* (I didn't see it anywhere else). The Democratic National Committee decided "to include gays in the list of groups qualifying for affirmative action status in party conventions and on committees. . . . Now the party must actively locate gays to be delegates to future conventions." One wonders whether the press will find it possible to ignore this bombshell in 1988, if the same rule still obtains.

Under the rubric of "dignity," a codeword almost always encompassing undignified behavior these days, the Democratic platform further advanced the cause of homosexuals by pledging: "We will assure that sexual orientation *per se* does not serve as a bar to participation in the military."

But the award goes to Walter Mondale for making the one utterly forthright promise of the convention—the promise to raise taxes if elected. This amazing promise produced a stunned silence of perhaps half a second, followed by a tepid round of applause. My immediate reaction was that Ronald Reagan had just won the election. I also reflected that Mondale was perhaps half right when he said that the Republicans will do the same thing, because some of Reagan's top aides, notable James Baker, have been trying to get the President to approve another tax increase.

Mondale's promise to raise taxes if elected looked like a calamitous mistake for the Democrats. It was as though he and his advisers had read all those hundreds of editorials in *The Washington Post* and *The New York Times* urging "deficit reduction measures"—code phrase for a tax increase—and took the message personally when it was intended for President Reagan. The inside-the-beltway crowd had carefully prepared and disguised this elephant trap and found in San Francisco that they had caught . . . a donkey!

Well, that is enough for one convention. San Francisco is certainly a beautiful city. The temperature dropped down to the 50s in the evenings, which made up for a lot of hot air. One afternoon I went across the bay to Berkeley, to see what those notoriously radical students were up to. These were the first four bulletins I saw on the Student Notices board directly opposite Sproul Hall, the focal point of student activity at the university in the mid-1960s.

"MOVE WITH US. AEROBICS, STRETCHING, TONING.
TRY YOUR FREE INTRODUCTORY CLASS ANY MONDAY,
WEDNESDAY & FRIDAY AT AMERICAN BAPTIST SEMINARY."

"POWERFUL! HSING I. THE WORLD'S OLDEST INTERNAL
MARTIAL ART HAS IMPORTANT BENEFITS FOR TODAY.
GREAT FOR YOUR MIND!"

"T'AI CHI CH'UAN. THE JOY OF MOVING.
NEW CLASSES, AUGUST 1 AND 2."

"EMPTY GATE ZEN CENTER OF BERKELEY ANNOUNCES
A VISIT BY MASTER DHARMA TEACHER MU DEUNG SU NIM
FREE PUBLIC TALK, WEDNESDAY JULY 18."

Don't ask me what it all means. The students are said to be "searching," "looking for new meanings." Perhaps they've decided to give up on liberalism, although I'm sure their teachers haven't.

I walked down Telegraph Avenue, and the first book store I came to had just about everything by C.S. Lewis, Dorothy L. Sayers, Charles Williams, J.R.R. Tolkien and George McDonald. It was a nice summer afternoon, and everything seemed very peaceful by comparison with the Democratic National Convention. I would gladly have stayed there for the rest of the week.

Coronation in Dallas

Downtown Dallas is a bit of a mystery to me. I hadn't been there for fourteen years. It is about a mile square, and seems to consist of 50 or 60 skyscrapers set down in the midst of a great deal of Allright parking. Surrounding the skyscraper archipelago is an encircling reef of freeways. There seem to be few shops, restaurants, or even people, as far as I could see. Just a mass of offices piled on top of one another.

The weather was appalling. Have you ever opened an oven not knowing it was on, and then backed off from the wave of heat that hits you in the face? That was what it was like opening an outside door in Dallas at the time of the Republican National Convention. One baking hot Saturday afternoon I went to Neiman-Marcus, one of the few shops on Main Street. Outside the front door of the luxury store there stood an assortment of amateur preachers and winos, nearly all black. They were exchanging jibes and desultory comments back and forth as they waited for the very occasional bus to come trundling up the street. Every now and then a nervous-looking, well-dressed couple would emerge from the cool, dark shopping cave, only to run the gauntlet of proletarian inspection and comment.

"Let me baptize you!" called out a disheveled wino, his arms out horizontally as a young couple came out blinking into the Texas glare. They set off walking fast without looking back.

"Yeeuuoooo brother!" the wino called after the retreating pair. He hoisted a quart beer bottle to his lips and refreshed himself with a lengthy swig. "I got the spirit today," he said quietly, wiping his lips.

"Don't fight the word," said a young black preacher to shoppers, hesitantly emerging with their packages. He held a Bible next to his heart, and he wore navy blue pin-striped trousers and well-polished two-tone brown shoes. "The son of God kept the Sabbath! I know if you got the Holy Ghost."

174

I sensed there were other rich folk clustered just inside the doorway, anxiously peering out to see if the coast was clear. Across the street there was H. L. Green Co., "Cool Tomato Sandwich Shop." Someone had hung up a "Welcome Republicans" sign. Huge glass skyscrapers towered up behind in the merciless sun. The mostly deserted street was strewn with paper and leaflets.

"There's no downtown in Dallas," a Neiman-Marcus salesman told me inside the store.

"Business is good all the same," said another.

"When I came here 20 years ago," said the first, "I used to tell people back in New York that the streets in Dallas are so clean you could eat meals off them."

No one would repeat that old Dallas cliche today—not downtown, at least.

There has been tremendous growth in the suburbs, especially to the north along the LBJ Freeway, where a whole new city seemed to be under instantaneous construction. There you feel the real life of the city is concentrated, and there you can believe that Dallas is the fastest growing city in the country: lots of land, lots of opportunity, and lots of ready and willing labor from across the Rio Grande.

I wanted to go and inspect the tent city organized by a left-wing, U.S. Catholic Conference-funded outfit called Association of Community Organizations for Reform Now—ACORN. The Catholic bishops started picking up the ACORN check after the U.S. government—more specifically, ACTION director Tom Pauken— stopped funding it in 1981. Pauken then blew the whistle on the USCC's Campaign for Human Development—pointing out that it is in the business of funding leftist causes. The bishops responded by raising hell with the White House and trying to get Pauken, a Catholic, fired from his job. So far without success.

The ACORN protestors were encamped in the Trinity River bottoms, near Dealey Plaza (where President Kennedy was shot). Sorry, readers, the ACORN camp would have been well worth a visit, but it was just too hot for me that day, and so I decided instead to go and see some my solidly conservative Dallas friends, the Flints, who live a mile or so to the north.

"Phew," I said when I arrived, "is it normally this hot in August?"

"Ah known it hotter," said Mr. Flint, a knobbly old fellow who was walking about pigeon-toed in cowboy boots. He had once been a car salesman. The newspapers next day said it was 108 degrees at Love Field, a record.

He had been reading about the ACORN nuts and peace 'n' justice nasties camped out in Dallas, hoping to embarrass the Republicans.

"Reckon the rattlesnakes should get one or two of them," he reflected. The blinds were down and the air conditioner humming.

"Yew mean water moccasins, daddy," said his rhinestone-spectacled wife, who was coming in from the kitchen with a most welcome tray of Coors and Budweiser. Somehow I felt immensely cheered. America overall might be changing in ways that did not bode well, but Dallas still had some of the old spirit.

The next day I went to the famously wealthy First Baptist Church downtown, where a congregation of about 1200 was gathered. There was a 30-piece orchestra and a robed choir perhaps 200-strong. The church's pastor of 40 years, the Rev. A. W. Criswell, delivered a cogent sermon. Unfortunately I missed the earlier service, with the Rev. Jerry Falwell presiding. I suppose that hearing Falwell at First Baptist is the contemporary American equivalent of the 17th century English enthusiasm for turning out to hear the Puritan divines in their pulpits.

Falwell and Criswell—an embarrassment of Baptist riches! Criswell has the face of a 19th century wagon train leader heading into Indian territory. He spoke fluently, without notes, for perhaps 20 minutes, touching on by now familiar themes: the secular humanism that "tears our schools apart"; Communism; "the awesome inroads of sin that we never named before"; and our "enemies with great giant nuclear missiles pointed toward the city of Dallas"— this more a call to repentance than to higher defense spending.

First Baptist inside and out looked to be uncontested Reagan country. (Criswell in fact delivered the final benediction at the Republican convention.) A voter registration table was set up outside the church, and in a window I saw a sign saying "Reagan Bible Class," whatever that might be.

Well, I should hurry on to the convention, which I know you are dying to read about one more time. To tell the truth, it wasn't very thrilling. It *was* a sort of coronation. The convention center was much better than the one in San Francisco, bigger and cooler and more suitable, and with even more press freebies, shamelessly accepted all 'round.

For the delegates it was pretty much a working holiday. Remember, they were mostly appointed, not elected, because there weren't any contested Republican primaries. So these were people who had become delegates by virtue of their connections with the Republican hierarchy, not thanks to the voters. That is why they were mostly "moderates"— George Bush supporters. Sixty-two percent of them supported a nuclear freeze, according to an oft-quoted *Los Angeles Times* poll. So much for the Republicans' "right-wing" convention.

I saw something of the Indiana delegation because they were bivouacked at the same hotel. I was impressed by how seriously they took their task, in view of its foreordained outcome. At breakfast on the first day they were earnestly going through an inch-thick briefing book, color-coded and thumb-indexed, to make sure that they didn't miss out on any meetings or caucuses.

I asked one prominent-looking member of the delegation whom he would support for the nomination in 1988. He, along with about half of the delegation, had been appointed by the party's state central committee. Well, he said, he was for George Bush (as were about 50 percent of all the delegates at the convention). Why? Because he had met him at an Elks Club cocktail party in Indiana a few years back, and George had seemed like a pretty good fellow. "I think he's got good credentials," my man said.

That, I think, typifies the kind of suport that Bush enjoyed at the convention, and enjoys within the party now. It is real enough, I suppose, but is likely to prove lukewarm, because it has so little to do with the issues. That, no doubt, is why Bush went out of his way to attack Richard Viguerie and other conservatives when he came to Dallas. He senses, correctly, that as long as there persists a wing of the party that really believes in something, it is likely to prove more influential than the managerial rump that he represents. Bush and his managers are merely the executors of an agenda set by others. I suspect the truth is that they would prefer these "others" to be moderate Democrats than conservative Republicans.

The "moderate" Republicans that Bush represents are the analogue of Tory "wets" in England. Mainly they live in dread that someone will take away their inherited privileges, and they are willing to abandon almost all principle as the price for maintaining their position in the party. (Here you have the true meaning of "pragmatism.") Republican moderates view the Republican Party itself as a font of privilege—a very nice, cozy club with jobs, limousines, honors, paychecks, plush seats in the second row of the stalls, and "loyal opposition" status. In sum, it provides them with the opportunity to be a part of the action in Washington, to be respected, and to live comfortably.

All these benefits they will be able to enjoy —*provided* (as Howard Phillips of the Conservative Caucus puts it) they are willing to lose gracefully where it counts—in the legislative arena. Moderate Republicans insist only that they be permitted to lose slowly, without too conspicuous a show of surrender. Having lost on the issues, moderate Republicans like Bush (and literally hundreds of others—both Howard and James Baker come to mind) see themselves as playing a harmonizing, peacemaking role in American politics by (as it were) offering bipartisan resolutions to make politics unanimous, thereby cementing in place any advance in the liberal agenda.

The one thing that threatens this cozy, pragmatic, Republican defeatism is the existence of those dreaded ideologues, people like Howard Phillips and Richard Viguerie and maybe Jack Kemp and of course other hardliners from North Carolina too horrible to mention, who are not content passively to play the good loser role. These people actually want to change things—troublemakers, as the moderates see them. They're not content to watch the play from the stalls, but actually want to rewrite the script (as though that were possible). Reagan, himself a closet moderate, has said that the right wants to march off the cliff, flags flying. (Don't these unrealistic hardliners realize that Republicans in this day and age have to be a little bit careful?)

So that's why Bush and the Bakers and the Doledrums don't like the conservatives. But my sense is that as long as the Republican Party continues to vote for its presidential candidates (in primaries), these people are not likely to do very well at the grassroots level.

On the first day I went to hear Jeane Kirkpatrick speak. She is not a "moderate," but then she is not a Republican either. I liked what she had to say, and she was well received by the delegates, but I am not sure that she came across on TV all that well. Jeane still has some fight left in her, unlike your typical Republican representative of corporate America. Her repudiation of the "San Francisco Democrats" (as a party *lately* gone haywire) was much better than Barry Goldwater's wretched reference to "Democrat wars" (implying a party permanently at fault). Kirkpatrick's comments were calculated to win Republican converts: Goldwater's to dissipate Republican strength.

One day I heard M. Stanton Evans say in his droll voice (no moderate he, by the way): "What the Republicans need is some orators." Maybe. But once again, fiery speeches presuppose belief in something.

Undistinguished as they were, the Republican speakers sounded considerably more conservative in Dallas than they do within the Washington D.C. beltway. One is reminded of Senator Dole's cynical comment that the GOP platform (supposedly so conservative but actually less so than in 1980, according to Howard Phillips), would be "forgotten in five minutes." In other words, it was a document for Dallas, not Washington; for voters, but not for government.

Even Gerry Ford was heard to make one or two pro-Reagan utterances from the podium. This time last year he supported the Washington consensus position of higher taxes and less defense spending. Bob Dole, no less, was heard to extol the achievements of Reaganomics, even though in Washington he has been putty in the hands of the Tax Increase Consensus.

Well, I certainly didn't want to spend a whole evening listening to Gerry Ford and the Doledrums (Transportation Secretary Elizabeth Dole was also on the program), so I eagerly accepted when Tom Pauken asked me of I would like to go with him to a party given by Nelson Bunker Hunt at his ranch about 30 miles away. Bunker is one of the sons of the legendary Texas billionaire, H. L. Hunt.

Somewhere to the north of Fort Worth, it seemed to be. I vaguely imagined when we drove out there that sitting around a table would be: Pauken, Bunker and myself, with maybe a family retainer or two and one or two others from some inner circle of true belief. There turned out to be 2,000 guests at the Circle T Ranch. Good old Bunker was laying on quite a party for the National Conservative Political Action Committee.

It was a sight to see, by the light of the setting sun. There were steers and stage coaches, museum-piece Conestoga wagons, Mercedes and Lincolns, well-watered pasture, white picket fences, and everyone milling about out of doors amidst the open-air booths and kitchens. Masses of cooks and waiters served Mexican-style hors d'oeuvres on *china* plates and cocktails in great chunky crystal glasses—no plastic or paper cups here, folks. Even though it was still plenty hot most of the men wore jackets and ties and kept them on. Texas can be much more formal than one might imagine.

I bumped into my friend Jon Utley. We had our usual discussion about tax rates in the Third World, and then, gesturing at the lavish scene around us— Cadillacs and cattle ranch—he said that this was how many Europeans imagined that people in America live, without realizing how unusual it is. In the eyes of the Left, of course, the very scarcity of it would give even greater grounds for condemning it. Thank God envy is not my vice. Someone said it is the only one of the seven deadly sins that gives no pleasure at all.

We were ushered into an absolutely enormous tent, or marquee as they would say in Europe, for dinner. It was so well air-conditioned that you really needed your jacket. There must have been 250 tables, each with linen napkins and several bottles of wine; and so many waiters on hand that you were served within a few seconds of sitting down.

For some reason the media had been admitted—such a mistake. Someone had calculated, no doubt erroneously, that they could be wooed by kindness. The media act constantly to inflame and arouse envy, and without new opportunities to play that role many journalists would feel quite bereft of purpose. At one end of the tent there was a sudden flurry and glare of spotlights: Nelson Bunker Hunt had put in an appearance for the cameras. I went over to inspect and couldn't help noticing how placid and calm he looked as he was quizzed by the florid, perspiring Douglas Kiker.

His name tag read "N. Bunker Hunt," and he seemed like a very unassuming fellow to me, especially a few minutes later when he came blundering back through the half-dark tent, tripping over chair legs and sticking out his hand to introduce himself most disarmingly to total strangers.

Someone told me that Hunt personally lost $1.5 billion in the silver crash of early 1980. But this in no way mitigates the fury of those who accuse him of having "manipulated" the silver market. I believe the Commodities Futures Trading Commission actually responded to the rising silver price by raising margin requirements, thereby precipitating a selling wave. In effect the CFTC changed the trading rules in the middle of the game.

Many celebrities were on hand: blessing by Jerry Falwell; master of ceremonies Pat Boone; rounds of applause for Charlton Heston, Wayne Newton, the Osmond Brothers, and various Dallas Cowboys and Olympic gold medal winners. There were so many senators and congressmen that they weren't even introduced. Someone said we should call on them to "pass the prayer amendment, vote the B-1 bomber and get on with it."

Bob Hope provided the main entertainment, and it was a pleasure to see him in such good form. He is 81 years old, someone said. One gathered that the somewhat conservative tone to the proceedings did not greatly bother him. Ronald Reagan, he said at one point, "wears his hearing aid in his right ear because he doesn't want to hear from his left, and I don't blame him." Bunker Hunt, he added, "owns so much silver that he won't die, he'll just tarnish."

Several people were to be heard saying that the media coverage of the convention dispensed with all pretense of impartiality. The TV networks in particular seemed blatantly ideologized, with their endless comments on how "right wing" the GOP platform was. I can recall no TV comment on the Democratic platform—a document which gave the green light to sodomy.

The Dallas Times Herald resorted almost daily to low-grade editorial sarcasm against the Republicans. While we were in town at least, the paper featured a jeering squad of leftist columnists (Molly Ivins, ringleader), seemingly guided by the belief that biting the hand that feeds you is a sign of enlightenment. In its current phase the D.T.H. more than confirms one's suspicions that the Left now has few arguments, only ridicule and accusation. In its lead editorial on the first day of the convention, the paper chided the GOP's "flat worlders" who don't realize that "most Americans believe that the world is round."

A quite astonishing amount of ink was wasted on the GOP's failure to endorse the Equal Rights Amendment. At a party given for Jeane Kirkpatrick, I had an opportunity to discuss this with Phyllis Schlafly, my favorite activist on the political scene. For some reason she was wearing a medal. She deserves one, but I'm sure she had to pin it on herself. Everyone was writing about the ERA as though it were a live issue, she said. The problem was that the amendment had been killed in 1982, and if anyone wanted to revive it his remedy lay with a prominent Democrat, House Speaker Tip O'Neill, who in 1984 has failed to bring it up for a vote on the House floor. Constitutional amendments are *legislative* initiatives. Doesn't anyone in the women's liberationist camp know this? Doesn't Geraldine Ferraro? (In her acceptance speech, she said that if elected the Democrats would "pass" the ERA.)

On the other hand, I can understand why the Left keeps up the assault. A target that ducks and cowers only invites further bullying. The Republicans indeed do duck and cower when it comes to ERA, saying "the problem" should be "dealt with legislatively" (i.e., by statute). Thus they concede that there is a problem in need of remedy. Why don't they say: Men and women are not the same and we should drop the pretense that we are. As long as the Republicans continue to accept their adversaries' criticisms in principle ("the government today is taking care of more people than ever before . . ."; "we have appointed more women than ever before . . ."), then the Left can be expected to keep attacking. They would be foolish not to. The whole problem is that "moderate" Republicans are defined by their willingness to assent to the principles of their opponents. If they weren't they wouldn't be called moderates.

Quiz Time: Who wrote the following in a Dallas newspaper: "Even if Walter Mondale had not nominated Geraldine Ferraro as his running mate, this would be the year of women in presidential politics. An idea, you might say, whose time has come. Women, the largest voting bloc in America, will have a tremendous impact on the coming election. All of this spells trouble for the Republican Party. You know, the Gender Gap."

You thought maybe Ellen Goodman? Wrong, try again. In fact the author was *Dallas Morning News* (temporary) columnist Ronald Reagan, Jr. He concluded the column as follows: "Maureen [Reagan] thinks the party is moving in her direction, away from Phyllis Schlafly. I hope so. Meanwhile I'm proud to have a sister on the cutting edge of the Republican Party's enlightenment concerning women's issues."

Hats off, however, to the *Dallas Morning News*'s TV columnist, Ed Bark, who kept an eagle eye on the networks, particularly their treatment of that famous 18-minute film introducing Ronald Reagan, Sr. I quote:

> Chief political reporter Roger Mudd had a grand time playing shrink. "Tom," he said, "this film will not tax your mind, it will not challenge your intellect, but it will attack your emotions head on."
>
> How many times will we see the American flag? Twenty-six times, said Mudd.
>
> How many times will Mudd talk down to viewers by leading them by the hand? One more time.
>
> "It is a powerful piece of film," he counseled, "but it is nothing the viewer can't handle. In fact, it could have the opposite effect, and there could be viewers the film cannot handle."

Apparently we have reached the point in this country where our most famous newsmen are openly reluctant to expose their viewers to patriotism.

I was back in the hall when the President gave his acceptance speech. It went on a bit too long, and he spent too much time comparing his record with Jimmy Carter's. We all know that Reagan's has been better, and it is superfluous to hammer away at the point. On the other hand, he was right to go on the offensive against Mondale and the Democrats in other respects, especially their disgraceful comparison of Grenada and Afghanistan.

I worry, frankly, whether Reagan has much of an agenda for his second term. We will need forceful leadership to resist the relentless pressure from the Left. Reagan has conspicuous faults: he seems to hold his beliefs all too lightly, so that he can easily be talked out of them. And, actor that he is, he craves popularity—"box office." He can't bear to be thought the villain, and has shown himself susceptible to bullying by those who do so portray him. Nonetheless, it is likely that Reagan will go down as a great President. As a friend of mine put it later, we'll miss him when he's gone.

As far as I was concerned, the best moment of the convention came at the very end, when Ray Charles sang "America the Beautiful" and everyone joined in. What a great singer he is—and I have never heard him sound better. It was a beautiful moment. As another old friend of mine in the vast hall, a Democrat, said: It was enough to make a Republican out of you. I asked around later but no one said they had seen it on television. Roger Mudd, Dan Rather & Co. must have prudently kept it off the screen, suspecting that many of their viewers wouldn't have known how to handle their emotions. But the tears were rolling down my cheeks.

Darwin's Mistake

How do we come to have horses and tigers and things? There are at least a million species in existence today, according to the paleontologist George Gaylord Simpson, and for every one extant, perhaps a hundred are extinct. Such profusion! Such variety! How did it come about? The old answer was that they were created by God. But with the increasingly scientific temper of the eighteenth and nineteenth centuries, this explanation began to look insufficient. God was invisible, and so could not be part of any scientific explanation.

So an alternative explanation was proposed by a number of savants, among them Jean-Baptiste Lamarck and Erasmus Darwin: The various forms of life did not just appear (as at the tip of a magician's wand), but evolved by a process of gradual transformation. Horses came from something slightly less horselike, tigers from something slightly less tigerlike, and so on back, until finally, if you went back far enough in time, you would come to a primitive blob of life which itself got started (perhaps) by lightning striking the primeval soup.

"Either each species of crocodile has been specially created," said Thomas Henry Huxley, "or it has arisen out of some pre-existing form by the operation of natural causes. Choose your hypothesis. I have chosen mine."

That's all very well, replied more conservative thinkers. If all of this life got here by evolution from more primitive life, how did evolution occur? No answer was immediately forthcoming. Genesis prevailed. Then Charles Darwin (grandson of Erasmus) furnished what looked like the solution. He proposed the machinery of evolution, and claimed that it existed in nature. Natural selection, he called it.

His idea was accepted with great rapidity. Once stated it seemed only too obvious. The survival of the fittest—of course! Some types are fitter than others, and given the competition—the "struggle for existence"—

184

the fitter ones will survive to propagate their kind. And so animals, plants, all life in fact, will tend to get better and better. They would have to, with the fitter ones inevitably replacing those that are less fit. Nature itself, then, has evolving machinery built into it. "How extremely stupid not to have thought of that!" Huxley commented, after reading *The Origin of Species*. Huxley had coined the term *agnostic*, and he remained one. Meanwhile, the Genesis version didn't entirely fade away, but it inevitably took on a slightly superfluous air.

That was a little over a hundred years ago. By the time of the Darwin Centennial celebrations at the University of Chicago in 1959, Darwinism was triumphant. At a panel discussion, Sir Julian Huxley (grandson of Thomas Henry) affirmed that "the evolution of life is no longer a theory; it is a fact." He added sternly: "We do not intend to get bogged down in semantics and definitions." At about the same time, Sir Gavin de Beer of the British Museum remarked that if a layman sought to "impugn" Darwin's conclusions it must be the result of "ignorance or effrontery." Garrett Hardin of the California Institute of Technology asserted that anyone who did not honor Darwin "inevitably attracts the speculative psychiatric eye to himself." Sir Julian Huxley saw the need for "true belief."

So that was it, then. The whole matter was settled—as I assumed, and as I imagined most people must. Darwin had won. No doubt there were backward folk tucked away in the remoter valleys of Appalachia who still clung to their comforting beliefs, but they, of course, lacked education. Not everyone was enlightened—goodness knows the Scopes trial had proved that, if nothing else. And some of them still wouldn't let up, apparently—they were trying to change the textbooks and get the Bible back into biology. Well, there are always diehards.

So it was only casually, about a year ago, that I picked up a copy of *Darwin Retried*, a slim volume by one Norman Macbeth, a Harvard-trained lawyer. An odd field for a lawyer, certainly. But an endorsement on the cover by Karl Popper caught my eye. "I regard the book as . . . a really important contribution to the debate," Popper had written.

The debate? What debate? This interested me. I had studied philosophy, and in my undergraduate days Popper was regarded as one of the top philosophers— especially important for having set forth "rules" for discriminating between genuine and pseudo science. And Popper evidently thought there had been a "debate" worth mentioning. In his bibliography Macbeth listed a few articles that had appeared in academic philosophy journals in recent years and evidently were a part of this debate.

That was, as I say, a year ago, and by now I have read these articles and a good many others. In fact, I have spent a good portion of the last year familiarizing myself with this debate. It is surprising that so little word of it has leaked out, because it seems to have been one of the most important academic debates of the 1960s, and, as I see it, the conclusion is pretty staggering: Darwin's theory, I believe, is on the verge of collapse. In his famous book, *On The Origin of Species by Means of Natural Selection, or the Preservation of Favoured Races in the Struggle for Life,* Darwin made a mistake sufficiently serious to undermine his theory. And that mistake has only recently been recognized as such. The machinery of evolution that he supposedly discovered has been challenged, and it is beginning to look as though what he really discovered was nothing more than the Victorian propensity to believe in progress. At one point in his argument, Darwin was misled. I shall try to elucidate here precisely where Darwin went wrong.

What was it, then, that Darwin discovered? What was this mechanism of natural selection? Here it comes as a slight shock to learn that Darwin really didn't "discover" anything at all, certainly not in the same way that Kepler, for example, discovered the laws of planetary motion. *The Origin of Species* was not a demonstration but an argument—"one long argument," Darwin himself said at the end of the book—and natural selection was an idea, not a discovery. It was an idea that occurred to him in London in the late 1830s, which he then pondered in the Home Counties over the next twenty years. As we now know, several other thinkers came up with the same or a very similar idea at about the same time. The most famous of these was Alfred Russel Wallace, but there were others.

The British philosopher Herbert Spencer was one who came within a hair's breadth of the idea of natural selection, in an essay called "The Theory of Population" published in *The Westminster Review* seven years before *The Origin of Species* came out. In this article, Spencer used the phrase "the survival of the fittest" for the first time. Darwin then appropriated the phrase in the fifth edition of *The Origin of Species*, considering it an admirable summation of his argument. This argument was in fact an analogy, as follows:

While in his country retreat, Darwin spent a good deal of time with pigeon fanciers and animal breeders. He even bred pigeons himself. Of particular relevance to him was that breeders bred for certain characteristics (length of feather, length of wool, coloring), and that the offspring of the selected mates often tended to have the desired characteristic more abundantly, or more noticeably, than its parents. Thus, it could perhaps be said, a small amount of "evolution" had occurred between one generation and the next.

By analogy, then, the same process occurred in nature, Darwin thought. As he wrote in *The Origin of Species*: "How fleeting are the wishes of man! How short his time! and consequently how poor will his productions be, compared with those accumulated by nature during whole geological periods. Can we wonder, then, that nature's productions should be far 'truer' in character than man's productions?"

Just as the breeders selected those individuals best suited to the breeders' needs to be the parents of the next generation, so, Darwin argued, nature selected those organisms that were best fitted to survive the struggle for existence. In that way, evolution would inevitably occur. And so there it was: a sort of improving machine inevitably at work in nature, "daily and hourly scrutinizing," Darwin wrote, "silently and insensibly working . . . at the improvement of each organic being." In this way, Darwin thought, one type of organism could be transformed into another—for instance, he suggested, bears into whales. So that was how we came to have horses and tigers and things—by natural selection.

For quite some time Darwin's mechanism was not seriously examined, until the renowned geneticist T. H. Morgan, winner of the Nobel Prize for his work in mapping the chromosomes of fruit flies, suggested that the whole thing looked suspiciously like a tautology. "For it may appear little more than a truism," he wrote, "to state that the individuals that are the best adapted to survive have a better chance of surviving than those not so well adapted to survive."

The philosophical debate of the past ten to fifteen years has focused on precisely this point. The survival of the fittest? Any way of identifying the fittest other than by looking at the survivors? The preservation of "favored" races? Any way of identifying them other than by looking at the preserved ones? If not, then Darwin's theory is reduced from the status of scientific theory to that of tautology.

Philosophers have ranged on both sides of this critical question: are there criteria of fitness that are independent of survival? In one corner we have Darwin himself, who assumed that the answer was yes, and his supporters, prominent among them David Hull of the University of Wisconsin. In the other corner are those who say no, among whom may be listed A. G. N. Flew, A. R. Manser, and A. D. Barker. In a nutshell, here is how the debate has gone:

Darwin, as I say, just assumed that there really were independent criteria of fitness. For instance, it seemed obvious to him that extra speed would be useful for a wolf in an environment where prey was scarce, and only those wolves first on the scene of a kill would get enough to eat and, therefore, survive. David Hull has supported this line of reasoning, giving the analogous example of a creature that was better able than its mates to withstand dessication in an arid environment.

The riposte has been as follows: A mutation that enables a wolf to run faster than the pack only enables the wolf to survive better if it does, in fact, survive better. But such a mutation could also result in the wolf outrunning the pack a couple of times and getting first crack at the food, and then abruptly dropping dead of a heart attack, because the extra power in its legs placed an extra strain on its heart. Fitness must be identified with survival, because it is the overall animal that survives, or does not survive, not individual parts of it.

We don't have to worry too much about umpiring this dispute, because a look at the biology books shows us that the evolutionary biologists themselves, perhaps in anticipation of this criticism, retreated to a fortified position some time ago, and conceded that "the survival of the fittest" was in truth a tautology. Here is C. H. Waddington, a prominent geneticist, speaking at the aforementioned Darwin Centennial in Chicago:

"Natural selection, which was at first considered as though it were a hypothesis that was in need of experimental or observational confirmation turns out on closer inspection to be a tautology, a statement of an inevitable although previously unrecognized relation. It states that the fittest individuals in a population (defined as those which leave most offspring) will leave most offspring."

The admission that Darwin's theory of natural selection was tautological did not greatly bother the evolutionary theorists, however, because they had already taken the precaution of redefining natural selection to mean something quite different from what Darwin had in mind. Like the philosophical debate of the past decade, this remarkable development went largely unnoticed. In its new form, natural selection meant nothing more than that some organisms have more offspring than others: in the argot, differential reproduction. This indeed was an empirical fact about the world, not just something true by definition, as was the case that the fittest survive.

The bold act of redefining selection was made by the British statistician and geneticist R. A. Fisher in a widely heralded book called *The Genetical Theory of Natural Selection*. Moreover, by making certain assumptions about birth and death rates, and combining them with Mendelian

genetics, Fisher was able to quantify the resulting rates at which population ratios changed. This was called population genetics, and it brought
great happiness to the hearts of many biologists, because the mathematical formulae looked so deliciously scientific and seemed to enhance the
status of biology, making it more like physics. But here is what Waddington recently said about *this* development:

"The theory of neo-Darwinism is a theory of the evolution of the
population in respect to leaving offspring and not in respect to anything
else. . . . Everybody has it in the back of his mind that the animals that
leave the largest number of offspring are going to be those best adapted
also for eating peculiar vegetation, or something of this sort, but this is not
explicit in the theory. . . . There you do come to what is, in effect, a
vacuous statement: Natural selection is that some things leave more
offspring than others; and, you ask, which leave more offspring than
others; and it is those that leave more offspring, and there is nothing more
to it than that. The whole real guts of evolution—which is how do you
come to have horses and tigers and things—is outside the mathematical
theory."

Here, then, was the problem. Darwin's theory was supposed to have
answered this question about horses and tigers. They had gradually
developed, bit by bit, as it were, over the eons, through the good offices
of an agency called natural selection. But now, in its new incarnation,
natural selection was only able to explain how horses and tigers became
more (or less) numerous—that is, by "differential reproduction." This
failed to solve the question of how they came into existence in the first
place.

This was no good at all. As T. H. Morgan had remarked, with great
clarity: "Selection, then, has not produced anything new, but only more
of certain kinds of individuals. Evolution, however, means producing new
things, not more of what already exists."

One more quotation should be enough to convince most people that
Darwin's idea of natural selection was quietly abandoned, even by his
most ardent supporters, some years ago. The following comment, by the
geneticist H. J. Muller, another Nobel Prize winner, appeared in the
Proceedings of the American Philosophical Society in 1949. It represents
a direct admission by one of Darwin's greatest admirers that however we
come to have horses and tigers and things, it is not by natural selection.
"We have just seen," Muller wrote, "that if selection could be somehow
dispensed with, so that all variants survived and multiplied, the higher
forms would nevertheless have arisen."

I think it should now be abundantly clear that Darwin made a mistake in proposing his theory of natural selection, and it is fairly easy to detect the mistake. We have seen that what they theory so grievously lacks is a criterion of fitness that is independent of survival. If only there were some way of identifying the fittest beforehand, without always having to wait and see which ones survive, Darwin's theory would be testable rather than tautological.

But as almost everyone now seems to agree, fittest inevitably means "those that survive best." Why, then, did Darwin assume that there were independent criteria? And the answer is, because in the case of artificial selection, from which he worked by analogy, *there really are independent criteria*. Darwin went wrong in thinking that this aspect of his analogy was valid. In our sheep example, remember, long wool was the "desirable" feature—the independent criterion. The lambs of woolly parental sheep may possess this feature even more than their parents, and so be "more evolved"—more in the desired direction.

In nature, on the other hand, the offspring may differ from their parents in any direction whatsoever and be considered "more evolved" than their parents, provided only that they survive and leave offspring themselves. There is, then, no "selection" by nature at all. Nor does nature "act," as it is so often said to do in biology books. One organism may indeed be "fitter" than another from an evolutionary point of view, but the only event that determines this fitness is death (or infertility). This, of course, is not something which helps *create* the organism, but is something that terminates it. It occurs at the end, not the beginning of life.

Darwin seems to have made the mistake of just assuming that there were independent criteria of fitness because he lived in a society in which change was nearly always perceived as being for the good. R. C. Lewontin, Aggasiz Professor of Zoology at Harvard, has written on this point: "The bourgeois revolution not only established change as the characteristic element in the cosmos, but added direction and progress as well. A world in which a man could rise from humble origins must have seemed, to him at least, a good world. Change per se was a moral quality. In this light, Spencer's assertion that change *is* progress is not surprising." One may note also James D. Watson's remark in *The Double Helix* that "cultural traditions play major roles" in the development of science.

Lewontin goes on to point out that "the bourgeois revolution gave way to a period of consolidation, a period in which we find ourselves now."

Perhaps that is why only relatively recently has the concept of natural selection come under strong attack.

There is, in a way a remarkable conclusion to this brief history of natural selection. The idea started out as a way of explaining how one type of animal gradually changed into another, but then it was redefined to be an explanation of how a given type of animal became more numerous. But wasn't natural selection supposed to have a *creative* role? Darwin had thought so. This is how the evolutionary theorists reponded:

The geneticist Theodosius Dobzhansky compared natural selection to "a human activity such as composing or performing music." Sir Gavin de Beer described it as a "master of ceremonies." George Gaylord Simpson at one point likened selection to a poet, at another to a builder. Ernst Mayr, Lewontin's predecessor at Harvard, compared selection to a sculptor. Sir Julian Huxley topped them all, however, by comparing natural selection to William Shakespeare.

Life on Earth, initially thought to constitute a sort of *prima facie* case for a creator, was, as a result of Darwin's idea, evisioned merely as being the outcome of a process and a process that was, according to Dobzhansky, "blind, mechanical, automatic, impersonal," and, according to de Beer, was "wasteful, blind, and blundering." But as soon as these criticisms were leveled at natural selection, the "blind process" itself was compared to a poet, a composer, a sculptor, Shakespeare—to the very notion of creativity that the idea of natural selection had originally replaced. It is clear, I think, that there was something very, very wrong with such an idea.

I have not been surprised to read, therefore, in Lewontin's recent book *The Genetic Basis of Evolutionary Change* (1974), that in some of the latest evolutionary theories, "natural selection plays no role at all." Darwin, I suggest, is in the process of being discarded, but perhaps in deference to the venerable old gentleman resting comfortably in Westminster Abbey next to Sir Isaac Newton, it is being done as discreetly and gently as possible, with a minimum of publicity.

Agnostic Evolutionists

The first time I saw Colin Patterson was at the American Museum of Natural History in New York City, in the spring of 1983. He was in the office of Donn Rosen, a curator in the museum's Department of Ichthyology, which is the branch of zoology that deals with fishes. Patterson, a paleontologist specializing in fossil fishes, was staring through a binocular microscope at a slice of codfish. In his early fifties and balding, he was wearing black corduroys and a smoking jacket affair of the kind that I associate with the Sloane Square poets of the "angry young man" generation—the generation to which Patterson belongs by age, and perhaps by temperament. I would later spend time with him in London, at the British Museum of Natural History, where he is a senior paleontologist, and at Cambridge University, where we attended a lecture by the Harvard paleontologist Stephen Jay Gould.

Patterson often conveyed an impression of moody rebelliousness. He is authoritative, the kind of person others defer to in a discussion; he is habitually pessimistic; and he seemed not at all sanguine about his brushes with other scientists—encounters that by the late 1970s had become quite frequent. Those with whom Patterson has been arguing are mostly paleontologists and evolutionary biologists—researchers and academics who have devoted their careers and their lives to upholding and fine-tuning the ideas about the origins and the development of species introduced by Charles Darwin in the second half of the nineteenth century. Patterson, it seemed, was no longer sure he believed in evolutionary theory, and he was saying so. Or, perhaps more accurately, he was saying that evolutionists—like the creationists they periodically do battle with—are nothing more than believers themselves.

In 1978, Patterson wrote an introductory book called *Evolution*, which was published by the British Museum. A year later, he received a letter from Luther Sunderland, an electrical engineer in upstate New York and a creationist-activist, asking why *Evolution* did not include any "direct illustrations of evolutionary transitions." Patterson's reply included the following:

> You say I should at least "show a photo of the fossil from which each type of organism was derived." I will lay it on the line—there is not one such fossil for which one could make a watertight argument. The reason is that statements about ancestry and descent are not applicable in the fossil record. Is *Archaeopteryx* the ancestor of all birds? Perhaps yes, perhaps no: there is no way of answering the question. It is easy enough to make up stories of how one form gave rise to another, and to find reasons why the stages should be favoured by natural selection. But such stories are not part of science, for there is no way of putting them to the test.

By 1981, Patterson's doubts about evolutionary theory were finding their way to the public. A sentence in a brochure he wrote that year for the British Museum began: "If the theory of evolution is true. . . ." In the fall of 1981, Patterson addressed the Systematics Discussion Group at the American Museum of Natural History. Once a month, the group meets in an upstairs classroom at the museum, opposite the dinosaur exhibit hall. The audience in any given month is likely to be made up of museum staff, graduate students from nearby universities, and the occasional amateur like Norman Macbeth, the author of *Darwin Retried*. (Systematics is a science of classification; taxonomists working in systematics study the way taxonomic groups relate to one another in nature.) There may be no more than fifteen people on hand when the discussion focuses on, say, fossil rodent teeth; or there may be 150 or more when Richard C. Lewontin, the renowned geneticist and author, gives a talk on the meaning (if any) of adaptation in biology.

Patterson's address was titled "Evolutionism and Creationism." Patterson is not a creationist, but he had been trying to think like one as a sort of experiment. "It's true," he told his audience, "that for the last eighteen months or so I've been kicking around non-evolutionary or even anti-evolutionary ideas." He went on:

> I think always before in my life when I've got up to speak on a subject I've been confident of one thing—that I know more about it than anybody in the room, because I've worked on it. Well, this time it isn't true. I'm speaking on two subjects, evolutionism and creationism, and I believe it's true to say that I know nothing whatever about either of them.

One of the reasons I started taking this anti-evolutionary view, or let's call it a non-evolutionary view, was that last year I had a sudden realization. For over twenty years I had thought I was working on evolution in some way. One morning I woke up and something had happened in the night, and it struck me that I had been working on this stuff for more than twenty years, and there was not one thing I knew about it. It's quite a shock to learn that one can be so misled for so long. Either there was something wrong with me or there was something wrong with evolutionary theory. Naturally I know there is nothing wrong with me, so for the last few weeks I've tried putting a simple question to various people and groups.

Question is: Can you tell me anything you know about evolution? Any one thing, any one thing that is true?

In the public mind, challenges to Darwin's theory of evolution are associated with biblical creationists who periodically remove their children from schoolrooms where they are being taught that man evolved from monkeys. Most Americans know about the Scopes trial of 1925, in which a Tennessee high school teacher was fined $100 for teaching evolutionary theory. Four years ago there was the trial in San Diego in which Kelly Seagraves, director of Creation Science Research Center, unsuccessfully sued the state of California over regulations governing the teaching of evolution in California public schools. (Seagraves wanted science teachers to be required to mention pertinent passages from the Book of Genesis.) What most people do not know is that for much of this century, and especially in recent years, scientists have been fighting among themselves about Darwin and his ideas.

Scientists are largely responsible for keeping the public in the dark about these in-house arguments. When they see themselves as beleaguered by opponents outside the citadel of science, they tend to put their differences aside and unite to defeat the heathen. The layman sees only the closed ranks. At the moment, we can, if we listen hard enough, hear fresh murmurs of dissent within the scientific walls. These debates are more complicated, perhaps, than the old contest, Science v. Religion, but they are at least as interesting, and sometimes as heated.

One of the least publicized and least understood challenges to Darwin and the theory of evolution—and surely one of the more fascinating, in its sweep and rigor—involves a school of taxonomists called cladists. (A "clade" is a branch, from the Greek *klados*; "cladist" is pronounced with a long a.) Particularly interesting—vexing, evolutionary biologists would say (and do)—are those who toil in what is called transformed cladistics, and who might be thought of as agnostic evolutionists. Like many who have broken with a faith and challenged an orthodoxy, the transformed cladists are perhaps best defined by an opponent—in this case, the British biologist Beverly Halstead. Asked not long ago in a BBC interview what he thought of transformed cladistics, Halstead replied: "Well, I object to it! I mean, this is going back to Aristotle. It is not pre-Darwinian, it is Aristotelian. From Darwin's day to the present we've understood there's a time element; we've begun to understand evolution. What they are doing in transformed cladistics is to say, Let's forget about evolution, let's forget about process, let's simply consider pattern."

Since Darwin's time, biologists have been absorbed in process: Where did we come from? How did everything in nature become what it now is? How will things change in the future? The transformed cladists—they are sometimes called pattern cladists—are not concerned primarily with time or process. To understand why, it helps to know that they are trained in taxonomy; they are rigorous, scrupulous labelers. Their job as taxonomists is to discover and name the various groups found in nature—a task first assigned to Adam by God, according to Genesis—and put them into one category or another. Taxonomists try to determine not how groups came into existence but what groups exist, among both present-day and fossil organisms. To understand that cladists believe this knowledge must be acquired before ideas about process can be tested is to understand the natural tension that exists between taxonomists and evolutionary biologists.

The transformed cladists have escalated the battle. In the 1940s and 1950s, years which witnessed the growth of evolutionary biology, taxonomists allowed themselves what might be called a bit of artistic license. (They called it the new systematics.) This occurred in part, no doubt, because taxonomy had come to be thought of as dull and stuffy—particularly by evolutionists like Sir Julian Huxley (the grandson of Darwin's champion Thomas Henry Huxley), who believed it was high time to cease being "bogged down in semantics and definitions." Taxonomists, in other words, were regarded as bookkeepers and accountants in need of a little loosening up. In his 1959 book *Nature & Man's Fate*, Garrett Hardin, now a professor of human ecology at the University of California at Santa Barbara, quoted a zoologist as giving this advice: "Whoever wants to hold to firm rules should give up taxonomic work. Nature is much too disorderly for such a man."

The transformed cladists think otherwise, and have sought to reestablish taxonomic rigor. In doing so, they have come to think that it is the evolutionists who have the problem—the problem being slipshod methodology. Colin Patterson, perhaps the leading transformed cladist, has enunciated what might be regarded as the cladists' battle cry: "The concept of ancestry is not accessible by the tools we have." Patterson and his fellow cladists argue that a common ancestor can only be hypothesized, not identified in the fossil record. A group of people can be brought together for a family reunion on the basis of birth documents, tombstone inscriptions, and parish records—evidence of process, one might say. But in nature there are no parish records; there are only fossils. And a fossil, Patterson told me once, is a "mess on a rock." Time, change, process, evolution—none of this, the cladists argue, can be read from rocks.

What can be discerned in nature, according to the cladists, are patterns—relationships between things, not between eras. There can be no absolute tracing back. There can be no certainty about parent-offspring links. Only inferences can be drawn from fossils. To the cladists, the science of evolution is in large part a matter of faith—faith different, but not all *that* different, from that of the creationists.

"I really put my foot in it," Patterson told me that day I first met him nearly two years ago. We were in a restaurant on Columbus Avenue near the Museum of Natural History, and he was recalling the talk he had given eighteen months earlier to the systematics discussion group. "I compared evolution and creation and made a case that the two were equivalent. I was all fired up, and I said what I thought. I went through merry hell for about a year. Almost everybody except the people at the museum objected. Lots of academics wrote. Deluges of mail. 'Here we are trying to combat a political argument,' they said, 'and you give them ammunition!'"

He ordered something from the menu and said: "One has to live with one's colleagues. They hold the theory very dear. I found out that what you say will be taken in 'political' rather than rational terms."

Patterson told me that he regarded the theory of evolution as "often unnecessary" in biology. "In fact," he said, "they could do perfectly well without it." Nevertheless, he said, it was presented in textbooks as though it were "the unified field theory of biology," holding the whole subject together—and binding the profession to it. "Once something has that status," he said, "it becomes like religion."

The founding father of cladistics was an entomologist named Willi Hennig. Hennig was born in what is now East Germany and spent the bulk of his career there, studying and classifying flies. At some point in the mid-1960s (there is very little biographical information available about him) he turned up in West Germany; he died there at the age of 63, in 1976. His principal work is *Phylogenetic Systematics,* an updated version of which was translated into English and published in the United States in 1966 by the University of Illinois Press. It is a difficult book, and an enormously influential one. By the 1970s, as the prominent evolutionary biologist Ernst Mayr wrote in *The Growth of Biological Thought,* a virtual Hennig cult had developed. A Willi Hennig Society was formed in 1980, and its fourth annual meeting, held last summer in London, was attended by some 250 scientists from around the world. Last month, the society published the first issue of its new quarterly journal, *Cladistics.* According to David Hull, the philosopher of science (he was at the meeting too), "among evolutionary biologists, cladistics is what everyone is arguing about."

At the heart of cladistics are the concepts of "plesiomorphy" and "paraphyly." A characteristic, or trait, is said to be plesiomorphic if it is found in a group of organisms of more general scope than the specific group under consideration. For example, all primates have hair, but hair is a characteristic found in a more general class of creatures—mammals. What Hennig called the fallacy of plesiomorphy is the belief that a characteristic (like hair) identifies and helps to define a particular species or order of animal life when in fact it defines a broader group.

Hennig also objected to the still common practice in biology of identifying a group of organisms only by the absence of certain characteristics. (His reasoning was Aristotelian; in *On the Parts of Animals,* Aristotle wrote that "there can be no specific forms of a negation, of Featherless, for instance, or of Footless, as there are of Feathered or Footed.") It was the lack of precision that bothered Hennig: a feathered animal is one thing (a bird); a non-feathered animal is anything (except a bird). Groups in nature defined by an absence of characteristics Hennig called paraphyletic.

By calling attention to paraphyletic groups, Hennig helped revive the rigor taxonomy once prided itself on. Colin Patterson and other transformed cladists have moved on to examine—and call into question—the crucial role that paraphyletic groups play in evolutionary theory. Patterson points out that the ancestral groups of the Darwinian tradition are always paraphyletic; that is, they are defined by the absence of characteristics. In his 1981 talk at the Museum of Natural History, for example,

Patterson touched on the subject of invertebrates. Invertebrates make up one of the two general categories of animals. The negatively defined group comprises a huge diversity of animals, from the simplest single-cell protozoan to insects, clams, worms, and crabs. What brings this wide array of creatures under one heading is their shared lack of a backbone.

Cladists like Patterson have pointed out that the term *invertebrate* does not serve a scientific function; it is too nebulous, too inexact for that. It also accurately describes strawberries and chairs. But the term does serve a rhetorical function, the cladists maintain. It makes possible the claim, found in many textbooks, that "vertebrates evolved from invertebrates." According to the cladists, the last two words of the four word statement do not contain any information that is not asserted as factual by the first two; "vertebrates evolved" simply *means* that the first vertebrate had parents without backbones. The transformed cladists claim that "vertebrates evolved from invertebrates" is a disguised tautology.

In his museum talk, Patterson said that groups defined only by negative traits have "no existence in nature, and they cannot possibly convey knowledge, though they appear to when you first hear them." Evolutionary biologists maintain that negatively defined groups make sense and serve a purpose; they tend to accuse the cladists, as one writer recently did in the magazine *Science*, of engaging in "verbal legerdemain." But Patterson and his colleagues point their fingers back at the evolutionists. Patterson, for one, has called the paraphyletic groups "voids."

What evolutionary theory does, the cladists say, is make claims about something that cannot be demonstrated by studying fossils. They say that the "tree of life," is nothing more than a hypothesis.

Nor do they believe it will ever be anything more than that. When asked about this in an interview, Patterson said: "I don't think we shall ever have any access to any form of tree which we can call factual." He was then asked: "Do you believe it to be, then, no reality?" He replied: "Well, isn't it strange that this is what it comes to, that you have to ask me whether I believe it, as if it mattered whether I believe it or not. Yes, I do believe it. But in saying that, it is obvious it is faith."

Some cladists would like it if all the talk about evolution just quietly went away. Evolution is not important to their work. That work involves finding the positive and verifiable characteristics of the various species and determining how all these species fit together in the animal kingdom: what patterns exist in nature. They are interested in the distribution of characteristics in nature, not in how they came into existence.

I recently spent some time with two cladists on the staff of the Museum of Natural History. I first met with Gareth Nelson, who in 1982 was named chairman of the department of ichthyology. Nelson graduated from the University of Hawaii in 1966 and he joined the museum staff a year later. The walls of Nelson's office were lined with boxes of articles from scientific journals, and a large table was covered with papers and jars stuffed with small, silvery fish preserved in alcohol: anchovies. Nelson is just about the world's expert on anchovies, although he told me that the number of people studying them (three or four) is much smaller than the number of anchovy species (there are 150 known species, and Nelson believes there are many more undiscovered). This disparity between the magnitude of the scientific "problem" and the number of people working on it is a commonplace in biology. Most laymen think that the experts have pretty exhaustively studied the earth's biota, when they have barely scratched the surface.

Nelson put the issue of evolution this way: In order to understand what we actually know, we must first look at what it is that the evolutionists claim to know for certain. He said that if you turn to a widely used college text like Alfred Romer's *Vetebrate Paleontology*, published by the University of Chicago Press in 1966 and now in its third edition, you will find such statements as "mammals evolved from reptiles," and "birds are descended from reptiles." (Very rarely, at least in the current literature, will you find the claim that a given species evolved from another given species.) The trouble with general statements like "mammals evolved from reptiles," Nelson said, is that the "ancestral groups are taxonomic artifacts." These groups "do not have any characters that are unique," he said. "They do not have defining characters, and therefore they are not real groups."

I asked Nelson to name some of these "unreal" groups. He replied: invertebrates, fishes, reptiles, apes. But this does not by any means exhaust the list of negatively defined groups. Statements imputing ancestry to such groups have no real meaning, he said.

What about the fossil record, I asked Nelson. Don't we know that from the fossils that evolutionary theory is true? Like most people, I thought the natural history museums had pretty well worked out the fossil sequences, much as in an automobile museum you can find the "ancestors" of contemporary cars lined up in sequence: Thunderbird back to Model T.

"Usually with fossils all you find are a few nuts and bolts," Nelson said. "An odd piston ring, maybe, or different pieces of a carburetor that are spread out or piled on top of one another, but not in their correct arrangement."

He maintained that too much importance has been attached to fossils. "And it's easy to understand why," he said. "You put in all this effort studying them, and you get out a little bit. Therefore you are persuaded that that little bit must be very important. I can get ten times more information per unit with recent fishes. So if you put in all that effort on fossils, you are inclined to say that the information you get is worth ten times as much."

Nelson said it was quite common for paleontologists to go to all the trouble of digging up fossils without realizing that the animals in question were still walking about. (Think of spending months hunting for a book in used-book stores without realizing it was still in print.) "Say you dig up a 50-million-year-old beetle," he said. "It looks like it belongs to a certain family, but there may be 30,000 species in the family. What do you do? Go through all 30,000? No, you just give it an appropriate-sounding name, *Eocoleoptera*, say. If it is a species that has been in existence for 50 million years, somebody else will have to find that out, because you don't have enough time. You're out digging in the rocks, not poking through beetle collections in museums."

I asked him about anchovy fossils. How far back do they go? "Well," he said, "Lance Grande, who was a student here recently, studied that, and it turns out that all the fossils previously described as anchovies are not anchovies at all." (Grande is now an assistant curator in the department of geology at the Field Museum of Natural History in Chicago.) "In other words," Nelson said, "the people who described them did not do a very good job. So the fossil record of anchovies was reduced to zero. However, there was something in the British Museum that I think Colin Patterson told Grande about, something from the Miocene in Cyprus; maybe 10 million years old. And it turned out to be an anchovy—the only known fossil. It has not yet been described in detail, but there is information suggesting it is the same kind of animal we find inhabiting the Mediterranean today."

A week or two after I met with Nelson I spoke to Norman Platnick, a curator in the museum's entomology department and an expert on spiders. On my way to see him on the fifth floor, I was joined in the elevator by a couple of bald assistants who were wheeling along on a cart what looked like a dinosaur head. I was reminded that for a long time the museum had the wrong head on its brontosaurus. One of the few bits of conventional wisdom about paleontology is that entire animals can be reconstructed from scraps of bone. Paleontologists now repudiate the idea, first enunciated by the French anatomist Baron Cuvier in the early 1800s.

Steve Farris, a professor in the department of ecology and evolution at the State University of New York at Stony Brook and the president of the Hennig Society, told me that Cuvier erected a monument to his own error in the form of a cement statue of an iguanadon, now at the Crystal Palace outside London. "The animal that Cuvier imagined was four-footed and resembled a rhinoceros," Farris said. "The complete skeleton of the iguanadon is now known—the animal was bipedal, with a long tail." As for the idea that the relationship of early animals to present-day ones is well established, Farris said: "When they are writing for a general audience, a lot of paleontologists do try to give that impression."

Not far from the elevator I found Platnick's orderly office: spiders (dead) inside little labeled bottles; book-filled shelves; journal articles neatly stacked. It would seem that professional biologists spend at least as much time studying each other's work as they do the world around them.

Platnick, who is rather square-shaped and bearded, told me that when he was an undergraduate at a small Appalachian college, he would go along with his wife when she collected millipedes. "But I was a wretched millipede collector," he said. "When we arrived home, all I would have in my jars would be spiders." So he started to study them. Today he has a Ph.D. from Harvard, and he and Nelson are co-authors of a book recently published by Columbia University Press entitled *Systematics and Biogeography: Cladistics and Vicariance*.

Spiders, which go back to the Devonian period, 400 million years ago, belong to the class Arachnida and the phylum Arthropoda. They are among the invertebrates, in other words, and are not well preserved in the fossil record. About 35,000 species of spiders have been identified, Platnick said, "but there may be three times that many in the world." He thought there were perhaps four full-time systematists examining spiders in the United States, "and perhaps another dozen who teach at small colleges and do some research." There is an American Arachnological Society, with 475 members worldwide, some of them amateurs. They meet once a year and discuss scorpions and daddy-longlegs, as well as spiders.

"Most of the spiders I look at may have been looked at by two or three people in history," Platnick said, adding that he would most likely be dead before anyone looked at them again.

I asked Platnick what was known about spider phylogeny, or ancestry.

"Very little," he said. "We still don't know a hill of beans about that."
We certainly don't know what species the ancestor of the very first spider
was. All we know of such an animal is that it was *not* a spider. (Otherwise
the "first spider" would not have been first.) We don't even know of any
links in the (presumed) 400 million-year chain of spider ancestry.

"I do not *ever* say that this spider is ancestral to that one," Platnick said
firmly.

"Does anyone?"

"I do not know of a single case in the modern literature where it's
claimed that one spider is the ancestor of another."

Some spiders have been well preserved in amber. Even so, Platnick
said, "very few spider fossils have been so well preserved that you can put
a species name on them." After a pause he added: "You don't learn much
from fossils."

In view of Platnick's comments about our knowledge of spider ancestry, I was curious to know what what he thought of the following passage
from a well-known high school biology text, *Life: An Introduction to
Biology*, by George Gaylord Simpson and William S. Beck, first published in 1957 by Harcourt Brace Jovanovich and still in print.

> An animal is not classified as an arachnid because it has four or five pairs
> of legs rather than three. It is classified in the Arachnida because it has
> the same ancestry as other arachnids, and a different ancestry from
> insects over some hundreds of millions of years, as attested by all the
> varying characteristics of the two groups and by large numbers of fossil
> representatives of both.

At that he threw himself back in his chair, and burst out laughing.

In this passage, Simpson and Beck were practicing the verbal sleight of
hand that has been common in evolutionary biology since the 1940s. All
we know for sure is that there is a group of organisms (in this case spiders)
that are identifiable as a group because they have certain unique characteristics. They have spinnerets for spinning silk, for instance, and thus we
can say that all organisms with spinnerets are spiders. (They share other
unique features, too.)

If we want to explain *why* thousands of members of a group have
features uniquely in common, that is another matter entirely. We can, if
we like, posit a theoretical common ancestor, the ur-spider, which transmitted spider traits to all its descendants. That is precisely what Darwin
did in *The Origin of Species*. But Simpson and Beck do something very
different. They say that the composition of the class Arachnida was

determined by examining not the features of spiders but their *ancestral lines*. But no such pedigrees are known to science—not just with respect to spiders but with respect to *all* groups of organisms.

The point stressed by the cladists is this: Unless we know the taxonomic relationships of organisms—what makes each unique and different from the other—we cannot possibly guess at the ancestral relationships. Things in nature here and now must be ordered according to their taxonomic relationship before they can be placed in a family tree. Thus the speculations of evolutionists ("Do X and Y have a common ancestor?") must be subordinate to the findings of taxonomists ("X and Y have features not shared by anything else"). If fossils came with pedigrees attached, this laborious method of comparison would not be necessary; but of course they don't.

"Stephen Jay Gould does his work without bothering about cladistics, I assure you," Platnick said, citing a recent paper by Niels Bonde, a paleontologist at the University of Copenhagen. Platnick went on to say the "the literature is replete with such statements as 'fossil X is the ancestor of some other taxon,' when it has not even been shown that fossil X is the closest relative of that taxon." (By "closest relative" he means that the two taxa form a group having unique characteristics.) "This is seen most commonly in accounts of human paleontology, but it is by no means restricted to it," Platnick said.

One reason why many laymen readily accept evolution as fact is that they have seen the famous "horse sequence" reproduced in textbooks. The sequence, which shows a gradual increase in the size of the horse with time, is dear to the hearts of textbook writers, in large part because it is on display at the American Museum of natural History. But when Niles Eldredge, a curator in the department of invertebrates at the museum and co-author with Stephen Jay Gould of the "punctuated equilibria" theory of evolution (which says that organisms stay the same for millions of years, then change quickly rather than gradually, as Darwin believed), was asked about it once he said:

> There have been an awful lot of stories, some more imaginative than others, about what the nature of that history [of life] really is. The most famous example, still on exhibit downstairs, is the exhibit on horse evolution prepared perhaps fifty years ago. That has been presented as the literal truth in textbook after textbook. Now I think that that is lamentable, particularly when the people who propose those kinds of

stories may themselves be aware of the speculative nature of some of that stuff.

When I brought the subject up with Platnick, he said he thought horse fossils had not yet been properly classified, or even exhaustively studied. I wanted to know whether Platnick believed that evolution has occurred. He said he did, and that the evidence was to be found in the existing hierarchical structure of nature. All organisms can, as it were, be placed within an inter-nested set of "boxes." The box labelled "gazelles" fits in the larger box labeled "ungulates" (animals with hoofs), which fits inside the "mammals" box, which fits inside "tetrapods" (four-footed animals), which fits inside "vertebrates." The grand task of taxonomy, Platnick said, is to describe this hierarchical pattern precisely, and in particular to define the traits that delineate the boundaries of each "box."

Whether taxonomy will ever fill in all blanks in the pattern is a question Platnick cannot answer. One problem, he said, is the shortage of taxonomists. "Systematics," he said, "doesn't have the glamour to attract research funds." Research grants have increasingly gone to molecular and biochemical studies; the result is that support for taxonomy at many institutions has, he said, "withered away." This bothered Platnick. "I am fully prepared to stand up to any biologist who says evolutionary theory is more important, or more basic. Without the results of systematics there is nothing to be explained."

I wanted to find out what those on the other side—evolutionary biologists and paleontologists—had to say about what the cladists are saying. First I went to the bookshelf. In his 1969 book *The Triumph of the Darwinian Method* (recently reprinted by the University of Chicago Press), Michael T. Ghislin, one of Darwin's greatest admirers, seems to be taking on the cladists (or trying to) when he writes:

> Instead of finding patterns in nature and deciding that because of their conspicuousness they seem important, we discover the underlying mechanisms that impose order on natural phenomena, whether we see that order or not, and then derive the structure of our classification system from this understanding.

I next looked in *Hen's Teeth and Horse's Toes*, one of Gould's volumes of reprinted essays. "No debate in evolutionary biology has been more intense during the past decade than the challenges raised by cladistics against traditional schemes of classification," Gould writes. He is not sympathetic to cladistics ("its leading exponents in America are among the most contentious scientists I have ever encountered"), but in his essay "What, If Anything, Is a Zebra?" he admits that "behind the names and the nastiness lies an important set of principles." These he enunciates, only to repudiate. He acknowledges that a strict taxonomy would eliminate groups like apes and fishes. But when cladists go this far, "many biologists rebel, and rightly, I think." Like his Harvard colleague Edward O. Wilson, the Frank B. Baird Professor of Science, Gould opts for the "admittedly vague and qualitative, but not therefore unimportant notion of overall similarity" of form.

I decided it would be a good idea to talk with a scientist who believes strongly in evolutionary theory. Last May I traveled to Boston to meet with Richard C. Lewontin, a geneticist, a one-time president of the Society for the Study of Evolution, a well-known writer on science, and currently Agassiz Professor of Zoology at Harvard. I had seen a quote from Lewontin used as a chapter heading in a book titled *Science on Trial*, by Douglas Futuyma. The quote, as edited, read: "Evolution is fact, not theory. . . . Birds evolve from nonbirds, humans evolve from nonhumans."

Lewontin was uncharacteristically attired in a regulation white lab coat when I first saw him (instead of his usual blue work shirt). We talked a bit about his stand against biological determinism. Finally it was time to get around to the point of my visit. What about these claims: Evolution is fact; birds evolved from nonbirds, human from nonhumans? The cladists disapproved, I said.

He paused for a split second and said: "Those are very weak statements, I agree." Then he made one of the clearest statements about evolution I have heard. He said: "These statements flow simply from the assertion that all organisms have parents. It is an empirical claim, I think that all living organisms have living organisms as parents. The second empirical claim is that there was a time on earth when there were no mammals. Now, if you allow me those two claims as empirical, then the claim that mammals arose from non-mammals is simply a conclusion. It's the deduction from two empirical claims. But that's all I want to claim for it. You can't make the direct empirical statement that mammals arose from non-mammals."

Lewontin had made what seemed to me to be a deduction—a materialist's deduction. "The only problem is that it appears to be based on evidence derived from fossils," I said. "But the cladists say they don't really have that kind of information."

"Of course they don't," Lewontin said. "In fact, the stuff I've written on creationism, which isn't much, has always made that point. There is a vast weight of empirical evidence about the universe which says that unless you invoke supernatural causes, the birds could not have arisen from muck by any natural processes. Well, if the birds couldn't have arisen from muck by any natural processes, then they had to arise from non-birds. The only alternative is to say that they did arise from muck— because God's finger went out and touched that muck. That is to say, there was a non-natural process. And that's really where the action is. Either you think that complex organisms arose by non-natural phenomena, or you think that they arose by natural phenomena. If they arose by natural phenomena, they had to evolve. And that's all there is to it. And that's the only claim I'm making."

He reached for a copy of his 1982 book *Human Diversity*, and said: "Look, I'm a person who says in this book that we don't know anything about the ancestors of the human species." (He writes on page 163: "Despite the excited and optimistic claims that have been made by some paleontologists, no fossil hominid species can be established as our direct ancestor. . . .") "All the fossils which have been dug up and are claimed to be ancestors—we haven't the faintest idea whether they are ancestors. Because all you've got, and the cladists are right. . . ." He got up and began to do his famous rat-a-tat-tat with a piece of chalk on the blackboard. "All you've got is Homo sapiens there, you've got *that* fossil there, you've got another fossil *there* . . . this is time here . . . and it's up to you to draw the lines, because there *are* no lines. I don't think any one of them is likely to be the direct ancestor of the human species. But how would you know it's *that* [pat] one?

"The only way you can know that some fossil is the direct ancestor is that it's so human that it *is* human. There is a contradiction there. If it is different enough from humans to be interesting, then you don't know whether it's an ancestor or not. And if it's similar enough to be human, then it's not interesting."

He returned to his chair and looked out at the slanting rain. "So," he said, "look, we're not ever going to know what the direct ancestor is."

What struck me about Lewontin's argument was how much it depended on his premise that all organisms have parents. In a sense, his argument includes the assertion that evolutionary theory is true. Lewontin maintains that his premise is "empirical," but this is so only in the (admittedly important) sense that it has never to our knowledge been falsified. No one has ever found an organism that is known not to have parents, or a parent. This is the strongest evidence on behalf of evolution, although it is never stated as such.

* * *

Our belief, or "faith," that, as Patterson says, "all organisms have parents" ultimately derives from our acceptance of the philosophy of materialism. It is hard for us to understand (so long has materialism been the natural habitat of Western thought) that this philosophy was not always accepted. In one of his essays on natural history reprinted in *Ever Since Darwin*, Stephen Jay Gould suggests that Darwin delayed publishing his theory of evolution by natural selection because he was, perhaps unconsciously, waiting for the climate of materialism to become more firmly established. In his 1838 *M Notebook* Darwin wrote: "To avoid stating how far I believe in Materialism, say only that emotions, instincts, degrees of talent, which are hereditary are so because brain of child resembles parent stock." Darwin realized that the climate *had* changed—that evolution was "in the air"—in 1858 when he was jolted by Alfred Russel Wallace's paper outlining a theory of the mechanism of evolution very similar to his own.

The theory of evolution has never been falsified. On the other hand, it is also surely true that the positive evidence for evolution is very much weaker than most laymen imagine, and than many scientists want us to imagine. Perhaps, as Patterson says, that positive evidence is missing entirely. The human mind, alas, seems on the whole to find such uncertainty intolerable. Most people want certainty in one form (Darwin) or another (the Bible). Only evolutionary agnostics like Patterson and Nelson and the other cladists seem willing to live with doubt. And that, surely, is the only truly scientific outlook.

Culture of
the Cultured

CRETIN CRANKSHAFTS

One of my favorite spots in Washington is Gravelly Point, a small park located a couple of hundred yards north of the main runway at National Airport. You sit there on the grass and these huge jets come lumbering, then racing down the runway straight at you, noses tilted up, straining and roaring at the difficulty of getting off the ground, and finally ripping the atmosphere with a stupendous sound as they slice their way not very many feet over your head.

This is quite popular entertainment of a Sunday afternoon. The parking lot is always filled with cars and campers, the grass dotted with picnickers. Why it is allowed at all is to me one of the great mysteries of Washington. No doubt a plausible claim to danger could be asserted by one or another arm of government, thereby enabling the authorities to put the park off-bounds. But to date none has, and so it is possible to get a marvelous close-up of these big aircraft in motion. The thought crossed my mind the last time I was there that an hour at Gravelly point is more profitably spent than an afternoon at a museum of modern art.

It is hard to deny, yet it is rarely said, that the creative impulse was redirected at some point early in this century, or perhaps in the 19th, away from some of its normal artistic channels and into new ones associated with engineering and technology. Quite apart from being useful, a Boeing 747 is a far more impressive aesthetic object than most of what passes for "art" in our contemporary museums—the receptacles, as Vladimir Nabokov so well put it, of "crankshaft cretins of stainless steel, zen stereos, polystyrene stinkbirds, *objets trouves* in latrines. . . ."

Perhaps unconsciously, many modern sculptors pay imitative tribute to technology. Their cockeyed assemblages salute the unacknowledged creative realm of our century.

Bearing this in mind, I decided to pay a visit to the brand new East Building of the National Gallery of Art. This trapezoid-shaped, I. M. Pei-designed, marble marvel has been praised so many times in print that you will not hear any more hosannas from me. At the entrance stands a large slab of smooth bronze, apparently shaped by centuries of exposure to the elements. Perhaps retrieved from the foot of some Zimbabwean waterfall? Wrong. A catalogue explains that it came from the foundry of Henry Moore, that much-praised 20th century nihilist.

Inside, one enters a great open space with glass ceiling, indoor trees and plants, escalators, and not a painting in sight. Sure enough, there is a cretinous crankshaft on display, and, suspended from a wall, what appears to be the exploded innards if a town-hall clock. This turns out to be the creation of someone called Anthony Caro, who must take himself very seriously indeed.

The main museum space is perfect for large cocktail parties, and to judge by recent issues of *The Washington Post*'s "Style" section, parties will be the building's true purpose—Jackie Onassis in attendance, let us hope. It comes complete with cocktail party "talking points" such as the giant Calder mobile suspended from the ceiling. (I think it is the sheer childishness and "innocence" of these mobiles that we are expected to admire; admittedly values all too rarely associated with modern art.) But I find the airplanes of Charles Lindbergh and Chuck Yeager, similarly suspended at the Air and Space Museum across the Mall, a good deal more interesting.

I asked an attendant where the art was. Upstairs, she told me rather snappishly. I trudged up a huge staircase, past an enormous and really horrible canvas 13 yards long, consisting of nothing but ugly splotches of black paint, portentously entitled "Reconciliation Elegy." (But I have to hand it to the dauber—he must be quite a salesman.) I entered a room containing pieces of sculpture by one David Smith. To judge by his notices you would think he was a major figure in the history of Western art. This Smith visited abandoned factories in Italy, where "he found numerous sheets of steel, as well as tools and factory machinery which he incorporated into his sculpture. The rate at which Smith produced finished compositions while at Voltri was unprecedented in the history of sculpture . . ." (says an explanatory text on the wall).

The rapidity may be easily explained. In Smith's hands the idea of sculpture has been "deconstructed," or reduced to mere metallic assortments, imitating what may have once been found in a blacksmith's shop. Smith returned to his smithy roots, you might say. Like Calder's mobiles, these works are little more than whimsical parodies of technology.

I think this parodic display impresses the art crowd because the "artist" in his atelier thereby displays a sensibility superior to that of the engineer. Imitation may be the sincerest form of flattery, but parody is the surest demonstration of superiority. In like manner, the impoverished aristocrat mimicking the uncultured accents of the *nouveaux riches* who press ever closer to his dilapidated castle, both manages to amuse his guests and to reassure himself that he is, if poorer, a man of infinitely superior sensibility.

Down the hall was an utterly clownish exhibition called "The Subjects of the Artist," intended as "a *preliminary investigation* into the themes and subjects in abstract expressionism." (My italics. Note, again, the imitation of scientific method.) The exhibition consisted of a few rooms of paintings by De Kooning, Mark Rothko, Barnett Newman, and Robert Motherwell. What amused me were the photographs of these earnest gentlemen, sitting in solemn contemplation of their incomplete canvases—pondering, you are given to understand, creative problems, much as Albert Einstein pondered creative problems.

In the Rothko room (audacious nonsense—a series of horizontally-divided, two-tone canvases of little interest beyond the gullibility of the times in which we live) people were actually intimidated into . . . whispering . . . very soft sibilant little whispers from the nervous congregation. Readers of Tom Wolfe's *The Painted Word* will be amused to know that viewers in the Barnett Newman room were not even pretending to look at the canvases (blank except for the occasional vertical stripe) but were reading a prominently displayed "explanation" of them.

The original National Gallery was financed by Andrew Mellon. This addition to it was financed by his son, Paul Mellon. One is impressed that so much money has stayed in private hands. But beyond that, the new building strikes me as a waste of money. Andrew Mellon's construction of the National Gallery was a genuinely useful and democratic act of philanthropy. His son's building, when the pretension is stripped away, is primarily for the benefit of the Art Crowd, whose cocktail parties could have been held elsewhere. Paul Mellon might have more profitably and more aesthetically invested in aircraft construction.

MODULAR LETTUCE

This is a story about Modular Mondrian and Crispy Romaine Lettuce. You see, I've been back to that shining new East Wing of the National Gallery of Art—sometimes described as a morphological masterpiece, at other times as vibrant, dynamic, taut, timeless, free-form and sly. But my purpose on this occasion was to take a look at the National Gallery cafeteria.

My theory was and is that going to an art gallery, a modern one at least, is primarily an affirmation of status: a silent declaration of one's superiority to the bulging-at-the-waist bourgeoisie. Furthermore, there's nothing more status-y than the food we eat. What food is served at an art gallery, then? It is a fitting subject for inquiry.

To digress briefly: I have a friend in Washington who is a very successful social-climber, and hence eager to disassociate herself from her former middle-American way of life. She has done a good job of this, severing as many ties as possible. But she has an Achilles Heel: an occasional urge for a McDonald's hamburger complete with pickles and onion strands and ketchup and mustard oozing out all over. Sometimes she gives in to the urge and heads off to McDonald's-in-the-'burbs. But she is not *completely* indiscreet on these occasions. I mean, she didn't get where she is today, on the staff of a Major Paper, by disregarding protocol entirely. So she wears a disguise, sort of, even though she knows there's not much chance of meeting people-who-count out there in Beltsville. Still, she puts on sunglasses, raincoat with collar turned up, big floppy hat: Greta Garbo with the Big Mac Attack.

Back to the National Gallery. I entered the East Building, and as I did so I became aware that I was in an exciting public space, designed for the future. The indoor plants were embedded as usual in chips of tree bark. (why has this become so common?) A Noguchi sculpture loomed into view—all 8 ½ tons of it, complete with chisel marks on the stone. This exemplifies "the direct contact of man and matter," the master himself has explained. (Did no one tell him that we don't allow sexist language here?)

I made my way down an access-for-the-handicapped ramp, and wondered, Why no solar heating panels? I then caught sight of Motherwell's "Elegy": brooding, looming, protesting, anguished, textural and contextual. Soon there was light at the end of the tunnel, a brightly tumbling waterfall, at length the cafetaria: lots of tables and grouped around them squat-green-Cubist, proudly-plastic, designed chairs which brought a word to the tip of the tongue: Mmmmm-m-m-m-modular. Anyway, modern. And now the menu: Beef Stroganoff, Mushroom Quiche, Crisp Romaine Lettuce, Perrier Water at $1 a bottle, or, if you prefer, little bottles of red or white California wine, with wine glasses fixed in place over the top. So you put all this on your tray, you sit in your modular chair, you unseal your wine glass, pour the wine, *toy* with your quiche, which is quite a decent quiche, and as you watch the prettily tumbling waterfall you will soon have visions of Noguchi playgrounds, Martha Graham's dance company, and *entire environments*, like Detroit's Civic Center, complete with bike lanes and signs in Spanish welling up into your consciousness: in short, a more just society humanized by art.

PORGY AND BASS

I have a fellow feeling for Alistair Cooke. Like him I came to the United States from England, much attracted by the native jazz. But still, despite hearing his broadcasts for years, I retain in my mind only an indistinct impression of the gentleman. He seems to have achieved the odd status of cultural master of ceremonies on permanent loan from Great Britain; as it were a performing museum piece, who sees through the daily political hurly-burly, his gaze undistractedly fixed upon the finer things of life; an entrepreneur of culture and history; fine-liver, true observer, and, yes, "superbly civilized voice," as we learn from the dust jacket of *The Americans*, a collection of his radio talks broadcast by the BBC.

Civilized. Cooke to a T.

After explaining that he has "summered and autumned on the North Fork (of Long Island) for forty-two years," Cooke takes us on a characteristic tour:

"There is nowhere I know—not the Mediterranean or the Crimea, most certainly not California—where there is such a succulent haul of so many kinds of splendid eating fish. We are just at the point where the northern cold-water fish nibble at our shores and where the warm water fish abound. First for the gourmet, are the noble striped bass and the blue fish. Then the swordfish, and the flounder, and the lemon sole. But there are also other very tasty species which city people either don't know about or despise out of genteel ignorance. In the summer months the fat porgy is always mooching along the bed of the bay . . . Baked porgy is delicious and I simply have no idea why it never appears on restaurant menus."

Because, silly, you can't expect city people in their genteel ignorance to know any better. Not to put too fine a point on it, they are uncivilized. No doubt of good yeoman stock and all that, but too busy speeding about on expressways, going to ball games and worrying about ephemera like inflation and housing costs to grasp the finer things of life such as the cuisine of North Fork.

At Christmas, Cooke is in northern Vermont. (Thank goodness, nothing too commercial for him.) Here it's two fat geese a-honking, old-fashioned kitchen platters tottering, many pans a-bubbling, and much ado about venison. Cooke must indeed be descended from cooks. "I ought to say that I've had venison in farmhouses in Scotland and in lush restaurants in London and Paris," Cooke-the-tourer reminds us, "and with an immense to-do and gaudy promises of foods for the gods, in Texas."

But don't you know it was better in Vermont? A man of the world, Cooke has had his meals among the rich and powerful and the men of the soil. We who rely on his cultural judgment are reassured to find him more at home with Scottish crofters and silent Vermont farmers (with homemade furniture in their parlors) than, say, the cast of "Dallas."

Well, soon enough it is time to return to New York City, to the eight-room rent-controlled ($900-a-month) apartment overlooking Frederick Law Olmstead's Central Park, which Cooke perceives as "a precious breathing space in a jungle of cement and steel."

Cooke looks out of his study window and notices the endless stream of endless joggers, concluding, after some side reflections on the arms race, that "earnestness is the only soil in which ideology can grow." Earnestness is a bore, ideology makes people behave dangerously.

All perfectly true. But Cooke's response, to assume the guise of the peripatetic aristocrat, devoted to little more than porgy and bass, is the intellectual equivalent of unilateral disarmament.

Cooke has plenty of talent, and I don't begrudge him his fame and fortune for a minute. You can't blame a man for being a success. He claims in an introductory note that his BBC talks are written "no more than a couple of hours before they are taped"—something that would be well beyond my powers.

From time to time, Cooke shows a fleeting impatience with the more obviously destructive developments in America since he arrived here in the 1930s. He alludes to pornography as "the clutter of filth that floats along with the First Amendment and is marketed for lucre in the name of liberty." And he includes in this collection a good piece on Richard Nixon, hinting here and there at a measure of dissent from his well-regimented, like-minded colleagues in the press. Cooke never writes about America in the condescending style of so many visitors from England.

On the whole, however, Alistair Cooke seems to recognize that to speak out too loudly about some of the more self-destructive tendencies in America would meet with disapproval in North Fork. The neighbors would take it amiss. So old Alistair pretty much minds his own business, not butting in too much, whistling his own quirky melody as he saunters down Fifth Avenue on his way to the recording studio. It would be nice to know what he really thinks. "Cooke's Jeremiad" would be perfect for public television. All the right people would be watching. Perhaps he could sing it as a swan song.

CRITIC AS BUILDING INSPECTOR

At first glance, the publication of Andrew Porter's arcane music criticism between hard covers seems puzzling. (*Music of Three Seasons*: *1974–77*.) Twenty dollars is a lot to pay for a collection of columns, originally published in *The New Yorker*. They were tedious enough when they first came out. The only excuse I can find for putting them in a book is that his publisher has discovered a subtle variation on the Vanity Press theme. A couple of thousand names are listed in the index, and I suppose a good many of these can be expected to pay their $20. As one might expect, Porter is suitably polite about the serious music scene in and around New York.

Earlier, Porter was the music critic of the London *Financial Times*. In everything that he writes he seems to be immensely pleased with himself. It takes some time for the reader to notice that he has very little to say about the actual music rather than last night's performance of it. This shortcoming is disguised with a good deal of scholarly and historical reference, smokescreens of trivia, lists of proper names, knowing allusions to texts. It pleases him to tell us that when the work was last performed (in Prague, say), the soprano used a corrupt text in the adagio, but today, at Tully Hall, she used a properly amended score. And this minor change gave him "rare pleasure."

Criticism of this type can scarcely be of interest to more than two or three people. The main point seems to be to show off one's own knowledge of such matters, thereby reducing a music column to a self-set trivia quiz.

One of the points that Porter loves to make in the course of a review is to lament that the lights were turned down during the performance, thereby preventing him from following the score or libretto.

This puts him in the Building Inspector school of criticism. He is the man who trudges about in the crypts and recesses of the cathedral, blueprints under his arm, making sure that the structure conforms in detail to the building specifications, worrying about the possibility of a hairline crack here or a piece of dislodged masonry there. Discrepancies are triumphantly noted in *The New Yorker*. Readers are not so much enlightened about music as they are reminded of Porter's credentials.

There is little evidence in this book that he is able to step back and simply enjoy the musical architecture. Perhaps he can. But if so, he can't write about it. Perhaps he regards such appreciative writing about music as a sign of amateurishness, hence beneath his calling as a professional and credentialled music inspector.

For example, Porter devotes a few pages to recent performances of Bach's *Goldberg Variations* and Beethoven's *Missa Solemnis*. These are two of the greatest works of Western music, as no doubt Porter would agree, but he has almost nothing to say about them: "What can be said about the *Goldberg* which is not summed up in Tovey's observations. . . ." (Quite a lot.) As for the *Missa Solemnis*, he tells us: "After the blaze of *Gloria in excelsis*, the composer's gaze drops to earth with the soft, low, *et in terra pax*." That's about as far as he will go in trying to "translate" the music's meaning into English.

Porter exemplifies the peculiar but widespread editorial   Porter exemplifies the peculiar but widespread editorial fear that criticism of the arts must be entrusted to certified building inspectors. Anyone else risks exposing the publication in question to a harrowing onslaught of expert obloquy. Editors have somehow been bullied into believing that only music "majors" are qualified to write about music.

In fact, 20th century writing about music strongly suggests that the best criticism comes from amateurs—George Bernard Shaw and Aldous Huxley come immediately to mind. And this is not entirely surprising, when one gives the matter some thought. The first thing to say about music is that it is ultimately quite mysterious. As Huxley himself put it, the sound that we hear is "appreciated" because we perceive it to be an "analogy" to something else. But what is it analogous to? It is very difficult to say, although when we are enjoying a piece of music, we do in some way "understand" what it is that the musical sounds are analogous to. Beethoven's music strongly conveys this sense of analogy or metaphor. The language of music that he uses so effectively "translates" the emotional content of some other world into our world of fugitive sounds.

A rediscription, in words, of the content of this "other" world should be the ultimate task of those who write about music. But it is rarely attempted, and as far as I know hardly ever considered at all by music professionals with higher degrees in the subject. J. W. N. Sullivan, a mathematician, did attempt something along these lines in his very interesting book on Beethoven—mainly a consideration of his late sonatas and quartets. One wishes more writers had followed in his footsteps. Writers and poets, who if they are any good have had some practice in the use of metaphor, at least have a chance of saying something interesting about music. Music school grads have very little chance. They know a lot about crochets and quavers—but literal crochets and quavers are exactly what we don't want to be told about. What is of interest is the meaning that the note-sequences convey. To attempt such an explanation entails shifting out of music-talk (Tovey-talk) into everyday English.

Twentieth century "classical" music has been dominated by two related developments, one musical and the other social. The first has been the complete breakdown of the very musical language with which earlier composers worked, using it to translate that "other" world into our world of sound. There simply is no longer any such language—merely an ever more absurd avant-garde Babel. Contemporary "classical" music, at least of the academic variety, no more deserves the name of music than a succession of words randomly chosen from the dictionary deserves the name of prose.

The second, social development has been the widespread pretense among academic critics such as Porter that this destruction of the musical tongue has not occurred. That, no doubt, is why the academy has been so anxious to appropriate to itself not merely the musical but also the critical function. There is an obvious possibility that if amateurs—mere music lovers—were given space to tell us what they think of modern music, they would not be afraid to tell us just how bad it is.

Much of modern academic music is little more than self-indulgent, subsidized solipsism. Whether he realizes it or not Andrew Porter lends a degree of legitimacy to this corrupt enterprise by having us believe that last night's erroneous detail really makes a difference; rather in the manner of the building inspector who finds a hairline crack—in a building in Hiroshima after the bomb was dropped.

Jonestown
Utopia

"He was trying to build a whole new kind of man in the jungle—the pure socialist man."

—ODELL RHODES

"There were people in that organization who gave up everything they had to follow a socialist dream."

—TERRI BUFORD

The death by cyanide poisoning of over 900 Americans in the Guyana jungle was surely the most dramatic and complete collapse of a utopian experiment ever recorded in history. And it was recorded, "live" on tape; and then later on film, dead. No failure of a utopian vision has ever been nearly so well documented or so swiftly promulgated before a worldwide audience. We may remain confident, nonetheless, that future proposals for "idealistic," collectivized, multiracial agrarian living will seem no less inviting as a result, nor will the number of candidates for such communal experiments in the eradication of individuality diminish.

There is much to be said about the collective death in Jonestown, but one of the most striking features is that the Rev. Jim Jones deluded his flock into following him into the jungle by telling them some very familiar lies about America. Ours is a "racist society" which held no future for them; a society to which it would be equally pointless to return because such deserters would be captured by the CIA or FBI in no time, at which point life would hardly be worth living. And so, in the last hour, a pathetic follower was reduced to writing to Jones: "Dad—I see no way out. . . . I

fear only that without you the world may not make it to Communism."
Larry Layton, who participated in the attack on Congressman Leo Ryan's
party at the Port Kaituma airstrip, later explained his actions by saying
that "I felt that these people were working in conjunction with CIA to
smear the People's Temple."

It is not surprising that Jones's semi-literate flock believed such lies.
They must have heard echoes of them a thousand times from their
television sets and seen them in newspaper headlines: a steady drumbeat
of hostility toward the CIA, for example, and constant talk of "racism"
from politicians on the make and from such columnists as Carl Rowan
who make a good living for themselves by portraying U.S. society in the
worst possible light. Jones did not invent these lies exactly. He merely
found them lying about, picked them up and adapted them to his
purpose, which was to persuade people to abandon their everyday lives
and submit to his will inside a concentration camp in the jungle.

Before reading about Jonestown, it's worth taking a look at *Nine Lies
About America*, by Arnold Beichman. There one may read that Dr. James
Cheek, the president of Howard University, said in 1971: "In 26 years
since waging a world war against the forces of tyranny and fascism and
genocide in Europe, we have become a nation more tyrannical, more
fascistic and more capable of genocide than was ever conceived or
thought possible two decades ago. We conquered Hitler, but we have
come to embrace Hitlerism." One wonders if any of the 700-odd blacks
who followed Jones to their deaths heard that lie from the lips of a leading
black educator before hearing it fleshed out by Jones.

In response to Jonestown, most media commentators have remained
conspicuously silent, perhaps only too painfully aware that Jones's advice
to his disciples had a distinctly familar ring to it, and must in some cases
have led them to wonder if Jones had been reading, clipping and filing
their own columns. Both Jones's diagnosis of American society, and his
socialist prescription for its remedy, decked out in vaguely "religious"
trimmings, have much in common with the outpourings of Colman
McCarthy in *The Washington Post*. Not surprising, then, to find
McCarthy nonplussed at the outcome of just the kind of collectivism he
professes to admire. Not even "profound explanations of deviant
behavior" could offer guidance in the Jonestown case, McCarthy felt.

The New York Times editorially had nothing to say, beyond cautioning
pre-emptively that "there can be no moral," a dubious assertion from a
quarter that has moralized about America in ways that must have seemed
very congenial to the Rev. Jones. In its "Week in Review" section of

November 26, 1978, the *Times* characterized Jones as having preached "a blend of fundamentalist Christianity and social activism," a distortion surely pleasing to those who worried that, in "explaining" Jonestown, it was important to shift the blame from Karl Marx to Jesus Christ.

One only had to turn to another page of that day's *Times* to read that the paper had in its files a 1977 interview with Jones's wife, in which she said that Jones "had not been lured to the ministry by deep religious faith, but because it served his goal of achieving social change through Marx." At 21, she said, "he decided that the way to achieve social change was to mobilize people through religion. 'Jim used religion to try to get some people out of the opiate of religion,' she said, adding that he had once slammed a Bible on a table and said, 'I've got to destroy this paper idol!'"

There were no religious services at Jonestown and subsequent reporting in *The New York Times* confirmed Mrs. Jones's observations. Documents in Jones's house contained lavish praise of Stalin and Mao Tse-tung, and a detailed description of how Jones figured out that religious trappings would serve to lure people toward his totalitarian goal. The *Times* also reported that a "tight clique of militant Marxists" surrounded Jones. Its reporting of the Jonestown massacre was voluminous and gave no impression of trying to camouflage Jones's true ideology. (It is opinion that is haywire at the *Times*.)

The New Yorker handed out its familiar little silver spoonful of genteel anti-Americanism, this being about the only political attitude that the magazine now considers safe. In William Pfaff's piffle, Jonestown was all a reflection of "the larger society . . . We have always lived with a good deal of violence in the United States. . . ." And so on.

The Washington Post had a man on the spot, Charles Krause, but the paper quickly recalled him so that he could contribute to a team-written paperback, *Guyana Massacre*. This book offers almost nothing in the way of guidance as to Jones's ideology, beyond noting that "he professed at times to be the spiritual heir of Christ and/or Lenin." This bland statement leaves the reader uncertain whether Jonestown exemplified Christianity or Leninism; but further perusal of the book would tend to resolve the reader's doubts in favor of the former, because nothing more is said about Jones's Communism while much is made of his "religion." ("His delusions, legitimized by his divine calling, had nothing to check them when he secluded his mission in the Guyana rain forest.")

In his reporting, Charles Krause actually seems eager to establish his own gullible nature, as though this were a desirable attribute in a *Washington Post* reporter. Even after being shot at and wounded in the hip, and even after writing on the front page of the *Post*, on November 23, that Jonestown was a "nightmarish concentration camp . . . its 800 to 1000 residents were kept prisoners by heavily armed guards and threats of death," Krause could at a later date write in the book: "As a reporter, I wonder now if I was terribly naive at Jonestown. Did I ignore evidence that Jonestown was closer to what the Concerned Relatives said it was—a concentration camp run by a madman—than the tropical paradise Jones and the others claimed? But that question is still not settled in my mind."

Krause had been so "impressed" during his visit to Jonestown with Congressman Ryan that he was even prepared to dismiss the attempt to stab Ryan as an "aberration." How he knew this is not clear, because wherever he went at Jonestown, "a reason was given why I really shouldn't wander around on my own. Someone would always come along and be friendly."

Although he only saw what he was allowed to see, Krause was eager to give Jonestown a favorable write-up in *The Washington Post*. "The truth was," he writes, "I had rather admired Jim Jones's goals." Then Jones's aides shot at Krause on the airstrip, and *still* he hasn't made up his mind about the place. Fidel Castro, who doesn't take kindly to foreign correspondents as a rule, might consider making an exception in Krause's case.

The Washington Post book reaches a low journalistic level in reporting Jones's attitude toward the Soviet Union. Chapter 14 begins as follows: "The myths, mysteries and primitive fears that surrounded the People's Temple in life, surrounded it in death. By one account Jim Jones, the cult's father, had escaped the holocaust at Jonestown and was headed for the United States with a squad of assassins to take revenge on his critics. Another story had lieutenants of Jones making their way out of the jungle with strongboxes of gold and diamonds and suitcases of money destined for the Soviet Embassy in Georgetown [Guyana]."

The reader is left to assume that both stories are equally absurd, because he knows that Jones has been definitely identified among the dead. But the second story was quite true (if one omits the "gold and diamonds") and *Washington Post* editors had every reason to know it was true. Considerable information about the Soviet connection was published in an Associated Press story printed (with a banner headline on the front page) in *The Washington Star* on November 27—well before the *Post* book was written.

But this AP story did not appear in the *Post*, thereby laying the groundwork for ridiculing it in the book. One gets the impression that everything at *The Washington Post* is in practice subordinated to the determination of its editors not to embarrass the Soviets. Weeks later, Krause found out what had been going on ("Cult Leader Earmarked $7 Million for Soviets"), only underlining his ignorance till then by noting that "this adds a curious new wrinkle to the Jonestown tragedy."

One other comment in a story by Krause was revealing. It is worth repeating even though others including Michael Novak and Reed Irvine of the AIM Report have drawn attention to it. "Asked if Jonestown had not been an experiment in fascism—with its armed guards and other means of preventing people from leaving—rather than an experiment in socialism, Jones [Jr.] replied: 'My father was the fascist. Jonestown was and still could be beautiful.' "

What we see in the reporting on Jonestown is just one more display of the great and utterly mysterious intellectual delusion of the 20th century, repeated over and over again: that the socialist doctrines of the revolutionary figures Jones explicitly admired (Lenin, Mao) are inherently noble, visionary and idealistic. If in practice they result in totalitarianism, some unforeseen ingredient has been added and one must call the result "fascism" (rather than allow the collectivist vision to be tainted).

But the truth is—and really one despairs that anything other than a small minority of college-educated people will ever see this—Marxist theory in practice is always totalitarian. And this is the great moral of Jonestown that *The New York Times* would not or could not see: Jones embraced Marxism and he proceeded to set up a Marxist "state" resembling in almost every particular Marxist states around the globe. Jonestown was the totalitarian regime in microcosm. Let us go over some of the details and note how familiar they will sound to refugees from the Soviet Union, Eastern Europe, China and Cuba:

The refusal to allow people to leave; the use of signed "confessions" to control "deviant" behavior; the lengthy meetings in the course of which ideological "error" (that is, the capacity for individual opinion) is eliminated; the repeated harangues over loudspeakers; the presence of informers, leading to a breakdown of trust among Jones's camp-followers turned prisoners; the censoring of mail; the contrast between egalitarian rhetoric and the privileged living conditions of those at the top (this has been excused by one or two loyal Jonestown survivors as "the necessity for structure"); the ready availability of brutal guards willing to enforce Jones's will. This readiness to exercise unlimited power over others must be assumed to be eternally present in human nature; which is why

Communism, based as it is on a hierarchy of power, will always degenerate into an exercise in brutality.

Perhaps most poignantly, in the case of Jonestown, there is the complaint by survivors that when potential rescuers arrived on Potemkin tours, no good ever came of it because those in authority always knew who was coming and so everything was carefully stage-managed and prettified beforehand. Several of these points, but particularly the last, are made with clarity and force in the remarkable "Affidavit of Deborah Layton Blakey," a girl in her twenties. This affidavit is printed as an appendix to the *Post* book; an appendix, but also an eloquent rebuke.

Finally, it is clear that Jonestown was like other totalitarian states in that there was no religious observance, veneration having been transferred from God to the head of state: Lenin in the Soviet Union, Mao in China, Castro in Cuba, Jones in Jonestown.

Theory and Practice
of the Hive

The metaphor of the bee hive to describe the world socialist enterprise was first used by Joseph Sobran in the summer of 1980, in his newspaper column, and at about the same time in *National Review*. When I first heard him talking about the Hive it struck me as a brilliant image, as it has ever since, greatly assisting our understanding of current events. In the first place, the metaphor gives a name to a phenomenon that for a generation has managed to elude labeling. If you don't have a name for something it is difficult to talk about it. Secondly, the image illuminates the key point that liberals and Communists belong to the same Hive. "Liberals don't feel they have any quarrel with Communism," Sobran wrote recently. "Why should one godless system oppose another?" Even to suggest such a thing today exposes one to the charge of "McCarthyism." The metaphor of the Hive enables one to say it anyway.

Finally, it shows the way around the great error that has for 30 years ruined earlier expositions of socialism.

The great mistake made by earlier students of the socialist network was to invoke conspiracy as its cause. Members of the John Birch Society, for example, noting the frequently coordinated or allied behavior of liberals, humanists, progressives, socialists and Communists, argued that there had to be a conspiracy somewhere in the background. Secret meetings were posited, in which everyone was given orders from Moscow. This was a serious mistake, as the liberals themselves were the first to know and delightedly point out. They knew that they were not taking orders from anyone. Thus they were able to make short shrift of the idea of conspiracy. Indeed, they could laugh it off. At the same time they ridiculed the very idea that there was a pattern whose origin needed to be explained. In this respect, however, the Birchers were correct. The pattern they had noticed was real enough. But they had misconstrued its source. Just because ducks fly south for the winter does not mean they first meet conspiratorially in Canada to discuss their flight plan. Likewise bees work in harmony, and different bees perform different tasks, but no bee takes orders from anyone. There is, however, a Queen Bee.

Over the years I have enjoyed many discussions with Joe Sobran about the Hive. He told me that he had independently worked out, from studying ACLU books, *The New York Times* and so on, that property, family and religion seemed to be the great targets of the Hive. Then, also in 1980, he read and brought to my attention a book called *The Socialist Phenomenon*, by Igor Shafarevich, a friend of Alexander Solzhenitsyn's and a Soviet mathematician at Moscow University. Having perused numerous documents dealing with the history of socialism that were available in the Moscow University library, Shafarevich concluded that it was a phenomenon that had endured throughout history, usually in the form of one or another Christian heresy. The destruction of private property, family and religion, and an obsession with material equality, were its perennial hallmarks, he concluded. Ultimately, he suggested, socialism aims to obliterate individuality and human nature—the human race itself. It embodies a death wish. In 1986, Joe Sobran told me that he thought "the goal of the Hive" was to "obliterate all traces of the West."

The Hive is such an immense subject that one hardly knows where to begin. It is a recent phenomenon. The Hive didn't exist in the 1940s, for example. But by the late 1960s it did exist. The Hive "evolved," and no one has yet traced that evolution. In an earlier, pre-Hive period there really was a Communist conspiracy. (That is why the Birchers have been so misled.) Read Whittaker Chambers's book *Witness*, or Allen Weinstein's *Perjury*, to see the way it worked. In pre-Hive times, a group of people with a conscious allegiance to Communism would meet in a room together, secretly, and pool information potentially helpful to the Communist cause. It might, for example, have been extracted from the State Department files by Alger Hiss. Whittaker Chambers would then courier it to the Soviet spy Bykov, "who maintained headquarters in New York City." Bykov would then take the information with him to Moscow.

Notice how slow, cumbersome, and even dangerous this mode of operation is. Above all it is based on a conscious loyalty to one system, and a conscious disloyalty to another; identity and definition are preserved.

But after World War II this system collapsed: Hiss was accused and convicted of perjury, the Fuchs spy ring exposed.

"The Left was crushed in America, quite suddenly," Sobran said. "It salvaged only one thing from this devastating era. It managed to label the era itself. While Stalin killed or enslaved millions, the Left named the era after a man who had, at worst, slandered dozens: Joe McCarthy. The 'fifties, the era of American supremacy, of booming capitalism, of family and faith, began a period of hibernation for the Left. It was winter."

Springtime came in the 1960s. The Old Left was succeeded by a new phenomenon: the Hive. Its ideology became inexplicit, unmentionable: a taboo. Labels were rejected. The old ones, we were repeatedly told, "don't mean anything any more." The very concepts of definition and identity came under attack. Loyalty was impugned—it came in the form of "oaths," which were portrayed as a McCarthyite relic. The new principles were not so much substantive as procedural: openness, dialogue, the people's right to know. At some point in the evolution of the Hive, documents that were secretly removed from State Department files, copied, returned and couriered to Moscow, were copied, returned and published in *Ramparts*, *The Washington Post*, and *The New York Times*. And this did not involve disloyalty—the old labels didn't mean anything any more. Did the people not have a right to know? Had one not read the Constitution and did the First Amendment not mean what it said? Was one proposing to curtail the Freedom of the Press?

The Hive has no Bible, but it has its doctrine. It has no written rules and no secrets, but its unwritten rules are hard to violate. It has no real headquarters, but the Queen Bee is in the Kremlin. It has no house organ, but its communication bees are numerous. It has no membership rolls, but it knows its enemies. It is not an organization, but it is organized. It is not a conspiracy, but it is capable of concerted action. It is intangible, but it is real enough.

The Hive is the progressive communion of all defectors, apostates, and heretics in our time. For those who have ceased to believe in the faith of their fathers or in the country of their upbringing, in flag, patriotism, judgment, heaven or hell, the Hive is waiting. Those who join the Hive find themselves at last part of an enterprise that they are comfortable with. They experience their abandonment of traditional values as liberation. There are no loyalty oaths in the Hive—but in practice the flight path of the bee is always deferential to the Queen.

The Hive is most visible when it is aroused. It was conspicuous, for example, when President Reagan said the Soviet Union was an "evil empire." When the Queen comes under attack the workers must drop all that they are doing and come to the defense of the Queen—by stinging the enemy. Reagan was repeatedly attacked by the Western media: bellicose, warmonger, insensitive, provocative, worst excesses of McCarthyism.

The Hive also fiercely disciplines errant bees. In 1982, Susan Sontag compared Communism to fascism—"fascism with a human face." The Hive didn't like that and the swarm came after her. "I have gotten so many grotesque attacks as a result of this," she told a *New York Times* reporter. "They're violent, sneering, and vituperative in a way which is very different from expressing strong disagreement. I'd never been the object of it before." Miss Sontag, meet Mr. Reagan.

As far as the Hive is concerned, your attitude toward the Soviet Union and Communism all but defines your status as friend or foe. You are not obliged to defend the Soviet Union, but you mustn't attack it. Someone like Anthony Lewis of *The New York Times* may do so from time to time, in a polite, constructive way. The Hive will take no notice. But he who makes a habit of attacking the Soviet Union, or Communism, will be treated as an enemy by the Hive and will be labelled a right-winger. The philosopher Sidney Hook, for example, calls himself a socialist, but he's strongly anti-Communist and so he's an enemy of the Hive. Calling himself a socialist doesn't make any difference. The label is ignored. Hook would attract no more criticism from the Hive if he were to call himself a capitalist.

Provided there is no immediate threat to the Queen Bee, the worker bees of the Hive feel free to pursue their daily tasks: attacking religion, family and private property in that order. To the extent that America protects property, nurtures families, encourages religion, the Hive is anti-American. Also, and more urgently, the Hive is anti-American because America has nuclear weapons, which make the Queen Bee uncomfortable. And now America threatens to defend itself against Soviet nuclear weapons, which is the last straw. The Hive knows that if such a defense were successfully deployed it would deprive the Queen of much of her power. Moscow is very unhappy about Strategic Defense. So, therefore, is the Hive.

The worker bees of the Hive are not conscious of "obeying" the Queen Bee, precisely because the Queen does not give orders. It merely makes its will felt. On May 15, 1987, for example, *Pravda* reported that Mikhail Gorbachev said the following in the course of a trip with Soviet military chiefs to Central Asia: "Now, as never before, it is imperative that world socialist movements and all progressive forces stand up to fight the forces which resist our disarmament efforts." Gorbachev said it was the Soviet "duty" to "show to the entire world the most serious danger of the SDI." His will is the progressive forces' command.

If there is a conflict between the interest of the Queen and the general Hive instinct to weaken property, family or religion, then the Queen's interest will always take precedence. This is perhaps the First Unwritten Rule of the Hive. For example, when the parents of 14-year-old Walter Polovchak decided that they wanted to return to the Soviet Union a few years back, they tried to take Walter with them. He, however, wanted to stay. Some were surprised that the American Civil Liberties Union then took the parents' side, using the ostensibly traditionalist argument that families should be unified. Around the country the ACLU came in for criticism for wanting to send young Walter back to the land of the Gulag. But, to the extent that it was and is the legal arm of the Hive, the ACLU knew what it was doing. It was acting in accordance with the First Unwritten Rule, even though doing so entailed taking a pro-family stance.

By examining other such Hive dilemmas, in which a choice must be made between conflicting Hive goals, it is possible to work out a Hive hierarchy of values. For example, there is commonly a Hive conflict between property and religion. A pro-lifer bombs an abortion clinic—an attack on property and therefore (one might guess) deserving of Hive commendation. Wrong! The anti-property act most probably has a religious motive. It will be fiercely condemned by the Hive—usually with the words, "despicable, cowardly." On such occasions the Hive will uncharacteristically defend property rights—showing Hive priorities: War on religion takes precedence over war on property.

In accordance with the First Law, the Hive will also defend property interests if this corresponds to the Soviet will. Armand Hammer, capitalist, is a friend of almost all Soviet rulers going back to Lenin—and today an authentic Hive hero. Does the Chevron Oil Company help the Cuban cause in Angola? Why then the Hive for once will side with Big Oil.

Faced with a conflict between weakening family and property, the greater Hive good consists in weakening family. The Hive supports free-and-easy divorce, which reduces marriage from the sacramental to the merely contractual state. The Hive is also staunch in its defense of the property rights of pornographers, contraceptive manufacturers, and condom producers.

* * *

To the extent that they are conscious of such matters (and most are not) Hive workers tend to take something like the following attitude toward the Soviet Union: She may not be pretty; in fact at times she can be pretty ugly; and yes she embarrasses us; but she's the only Queen we've got. The Soviets, after all, have the guns, and they have the will. They can if necessary keep the Sandinistas in the field, fighting imperialism indefinitely. Without the force and the will furnished by Moscow the whole progressive enterprise would very likely collapse. Like it or not, the Soviet Union is where the progressive principle was actually made flesh. Perfect it isn't. The Czarist tradition and the Russian mind may in the end prove to have been insuperable obstacles to the progressive vision. But the founding principle of the Soviet Union was correct. Does anyone have a better idea about how the New Society is to be established? If so, please speak up. If not, please temper your criticism. (By the way, how do you like this new Gorbachev? Not so bad, huh?)

Most Hive members are not conscious of any of this. They would be shocked, or more likely amused, to hear such sentiments imputed to them. Better to think of them, in fact, as Hive participants rather than "members." Joe Sobran likes to say that Jane Pauley of NBC's "Today" show is a more typical bee than, say, Anthony Lewis, or Richard Barnet of the Institute for Policy Studies, both of whom are not only on the left but know it. Jane Pauley, on the other hand, observes the progressive proprieties as automatically as she would table manners or good grooming. For her, as for millions of others, ideology is experienced as mere etiquette: as unwritten rules of behavior, the violation of which would be not so much erroneous as embarrassing.

During Ronald Reagan's 1980 campaign it was interesting to note that several times the communication bees of the media, posing as the arbiters of taste rather than the custodians of doctrine, exclaimed that Reagan had committed a "gaffe"—for example by saying that the progressive income tax was a Marxist idea (which of course it was, but that doesn't matter). Some things, you see, are simply not said. Not in polite society. To violate such rules is to breach etiquette—which is somehow less forgivable than to espouse wrong doctrine. It is one thing to proclaim oneself a Republican at the dinner table. It is quite another to burp at the dinner table.

As we know, Reagan before too long responded to the Hive's stings and arrows, dutifully saying he was sorry and promising never again to repeat the phrase "Evil Empire"; then proceeding to the arms control table, meeting with Gorbachev, and doing all those things that he had hitherto left undone and that the Hive had wanted him to do. Then, just as a nice new arms control agreement was on the horizon, who should butt in with some very unwelcome doubts but Henry Kissinger and . . . Richard Milhous Nixon!

James Reston, arbiter of decorum, wrote a *New York Times* column in the form of a letter to the offending parties: "Another point has to do not only with judgment but manners," he wrote. "If your proposals had been made to the President and the Secretary of State in private—an option always open to you—nobody could object, but to make them in public at a critical point in the talks is at least an act of discourtesy."

Bad policy, it turns out, is also in bad taste.

The worst of all breaches of etiquette is to notice the existence of the Hive itself, its constituent bees, and the predictable pattern of their various campaigns; that is, their affinity for the Queen Bee and their protective attitude toward the Kremlin. Pointing this out, gentlemen, is McCarthyism! In the worst of taste!

There is great irony here because the bees want to be able to recognize each other, and at the same time attack outsiders for recognizing them. "The mark of the bee is that he gets indignant if you call him a bee," Sobran said. "If a bee recognizes the Hive, that's as it should be. But if a non-bee recognizes the Hive, that makes him the enemy. It's really the whole principle of camouflage."

And herein lies a double irony. The affinity between Communism and liberalism was some time ago recognized by the Communists themselves. *They* have no compunction about appearing in parades, on platforms, and in various protest coalitions with every shade of liberal. Nonetheless, the liberals always turn a blind eye to their partners-in-coalition. And this enables them to treat the suggestion that they have an affinity with those to their left as one more right-wing delusion.

Even when it is pointed out to them directly, they are not embarrassed. In April, 1987, for example, there were well-publicized, coordinated Hive turnouts in Washington and San Francisco, the "Mobilization for Justice and Peace in Central America and southern Africa." AFL-CIO President Lane Kirkland warned his members about Communist manipulation and the participation of speakers "who are not committed to genuine trade union rights." The Newspaper Guild (representing journalists at such papers as *The Washington Post* and the news magazines) went right ahead and sponsored the event anyway. Told that the Communist Party, USA, was also a sponsor, David Eisen, director of information and research at the Guild said: "That does not bother us. No group should be excluded simply on the ground of its political position."

THE AD HOC HIKE

Tomorrow I'll be attending that well-advertized grievance jamboree, the "New Coalition of Conscience" on the Mall, celebrating the 20th anniversary of Martin Luther King's much publicized dream. Notice the word "coalition." It invariably points to an alliance of the Left. The march, or coalition, will be one more opportunity for various progressive groups to hector the nation with televised abuse and hostile slogans. It is predictable, incidentally, that the press will respect the Hive's unwritten rules of discretion in coverage. Probably without even giving the matter any thought, producers will assist the ideological goal of the coalition by decorously keeping the spotlight of attention away from all that is explicitly socialist.

What is Ruben Zamora, a Salvadoran "freedom fighter" (terrorist) doing at an American "jobs, peace and freedom" rally? Why has the U.S. Communist Party newspaper, the *Daily World,* given the march so much publicity, even telling readers "how to mobilize"? And why the participation of other Communist front groups, such as the National Lawyers' Guild and the U.S. Peace Council? If it were possible, the *Daily World* has been even more enthusiastic about the upcoming march than *The Washington Post.*

Media coverage to date has distracted attention from the reality of Communist *support* by playing up the false charge of Communist *control.* The ridiculousness of the latter spills over into the former, so that in the end any mention of Communism in connection with the march is embarrassing—not to mention McCarthyite. A few days ago, for example, *The Washington Post* ran a story ("March on Washington Called Communist Front") about a rally led by the Rev. Carl McIntyre, who "told 300 flag-waving, hymn-singing followers at the Washington Monument yesterday that next Saturday's march on Washington is a front for international Communism." By way of rebuttal, readers were told: "Donna Brazile, the march's national mobilization director, said McIntyre's charge is the result of 'paranoia' and that Communists played no role in organizing the march."

Liberals conveniently pretend that Communism is an all or nothing affair; that there are no degrees of it; that one is either "card-carrying" or completely unrelated to it; that the adjective "communistic," although used by Karl Marx, imputes "guilt by association," which is the essence of McCarthyism. But the truth is that those who knowingly associate with criminals are liable to prosecution as accessories. The media didn't hesitate to play up Secretary of Labor Ray Donovan's associations, and Frank Sinatra's. Can you imagine what a field day the liberals would have if there were an anti-Communist rally in Washington, with U.S. military and CIA figures, Richard Perle of the Pentagon, William Casey, Ed Meese and others "associating" on the platform with Ku Klux Klan members, John Birch Society members, Chilean police, Marcos cronies? Would Mr. Meese be able to deflect attention from this association by claiming that the rally wasn't under Klan *control*?

Clearly what the liberals have in mind is not that *association* is innocent, but that Communism is.

It is significant that both Communists and anti-Communists agree as to the "communistic" character of contemporary liberalism. The liberals themselves "reject labels" and put on airs as autonomous beings—free spirits. They actually play up the fact that the Right sees them as communistic. They proceed to "disavow" the different and false claim that they are Communist-controlled. At the same time they play *down* the Communists' perception of them as allies, for a very good reason: The Communists can reasonably be regarded as authoritative in such judgments.

You have to be wilfully blind not to see that the march's agenda is appealing to Communists. If implemented, defense spending would be greatly reduced; social spending and income-transfer programs would be greatly increased. Belief that jobs can be created by government edict would itself transfer even more power to Washington officials. To anyone who gives the matter a moment's thought it is obvious that the "Coalition of Conscience" would move America further down the road to socialism. And that is what it is intended to do. The anniversary of King's speech was only a pretext. As Atlanta mayor Andrew Young said on "Face the Nation" the next day: "There's no question that Ronald Reagan was the organizing factor that pulled this broad coalition together."

At Constitution Avenue I watched the Broad Coalition come rolling by— liberals, progressives, socialists, Trots and Communists: the Hive on parade. On one or two banners I noticed the words "ad hoc," a phrase which misleadingly emphasizes the accidental nature of such alliances. Most of the marchers were blacks who, one sensed, had dutifully responded to the various Hive appeals for a big turnout.

Muscular Communists came by toting a broadly coalitional banner, as wide as the road: "Communist Workers Party," it read, adding: "Self Determination for African Americans." I had thrust upon me an incredible amount of communistic literature, all filled with articles about the march. *The Guardian*, an "independent radical newspaper weekly," had a 20-page supplement on march themes. *The Daily World*, the Communist Party daily, had no less than ten articles on the march. The *Socialist Worker*, "paper of the International Socialist Organization," had six such articles. *El Salvador Alert* had two articles, and from this paper I learned that another coalition will be getting together on November 12, this one protesting "U.S. intervention" in Central America. In November, no doubt, Coretta Scott King will once again be participating, but as a peripheral member of the team rather than a central figurehead. Finally, *Worker's Vanguard*, the "Marxist Working Class Biweekly of the Spartacist League of the U.S." had three articles on the day's events.

The sheer volume of socialist literature that is published in America is staggering. A bookstore near Dupont Circle, Common Concerns, regularly stocks so many publications that there is no way their titles could be listed in the remaining space of this article. We are constantly told by *The Washington Post* and other publications that we live in a "conservative era," but I don't think many people can realize just how great is the disparity between the human energy that goes into promoting socialism and the energy that goes into resisting it. If the number of publications on either side are a guide, the forces of socialism outnumber the conservatives by about 100 to one.

It is true, of course, that at election time most voters will usually support the more conservative candidate, more or less instinctively. They won't necessarily even think of themselves as conservatives. They are, as the Left likes to say, "apathetic." The Left might with good reason hope they would stay that way. If the people ever were to become as actively involved in resisting socialism as the Hive is in promoting it the Hive's days would be numbered. The great asymmetry between Left and Right is that the former, believing that human nature urgently needs to be reformed, is ever active; while the latter, being basically content with life, is usually passive.

I couldn't help noticing at the march a fair number of those familiar old progressive reliables, tanned and lined couples now up into their seventies for the most part, grizzled and bearded, striding along in sensible ecological walking shoes. They have "college-educated" written all over them, and they're always white. They've been on the march since the 1930s, most likely. They look back with nostalgia to the time in 1965 when they saw the inside of a Tuscaloosa jail (if only for a few hours). She is wearing a beret, he some weird old fishing hat; both are plentifully bemedaled with buttons—the specific cause varying from month to month and march to march, but always in the same key; always in harmony with the Hive. Both are weighed down with rucksacks, filled no doubt with natural herbs, old copies of I. F. Stone's *Weekly*, a manifesto or two. They are truly imbued on these occasions with a spirit of selflessness. Onward they march toward the New Dawn, undaunted by old setbacks: the Gulag, the Castro calamity, Mao's murdered millions; sustained now by the vision of the Sandinista junta in Managua: New Managua Man!

As you watch them striding along you realize that they relish the outing, the ad hoc hike. Once more the communistic camaraderie! They wave cheerfully and familiarly to strangers—so diverse yet so reliably like-minded. Here a cap doffed to Concerned Catholics, there a hearty greeting to Quakers for Gay Civil Rights: all coming together, barriers crumbling, equality within reach. . . .

Ahead lay the Lincoln Memorial, and a familiar array of speakers who just happened to have been assembled for the occasion: Bella Abzug, Gloria Steinem, Harry Belafonte, Cora Weiss, Dick Gregory, Jesse Jackson, Benjamin Hooks, William Winpisinger; music by Pete Seeger, and Peter, Paul, and Mary.

I sat down on the grassy sward beside the Reflecting Pool, amidst numerous marchers who were unpacking their Free Spirit picnic containers and opening up their double six-pack Playmate igloo coolers. Out came the turkey sandwiches, the fried chicken wrapped in tinfoil, the clear plastic containers of potato salad, the cole slaw, sodas and brownies. Many of them had Panasonic radios glued to their ears, enjoying the peculiar pleasure of hearing about the event that they were part of.

We were so far from the platform that only the occasional word or phrase came wafting our way on the hot afternoon breezes: "Hunger in America . . . injustice . . . racism . . . regressive Reagan regime . . . committed to the elimination of Reaganism from the face of the Earth . . . fiscal austerity . . . struggle . . . visions . . . aspirations . . . Hitler." The black folks ate their fried chicken and sometimes said "Right on!" in solidarity. But mostly they went on picnicking.

"From across the nation," *The Washington Post* said the next day, "a diverse coalition of 250,000 Americans gathered at the Lincoln Memorial including 700 groups with a wide range of political and social agendas." No fewer than 21 *Post* reporters contributed to the paper's coverage, but the only word about Communist participation was a quotation from Harry Belafonte repudiating the idea that the march was "the result of some massive conspiracy."

On a Saturday TV talk-show, the columnist Jack Germond said he didn't "buy the idea that the march was a commie plot."

Not plot, Jack. Coalition.

Mitch Snyder's Crusade

It was a hot summer afternoon on Euclid Street and Mitch Snyder of the Community for Creative Non-Violence was working on a legal brief. Three or four cats were arrayed two-dimensionally on the wooden floors. A black youth in a white T-shirt was stretched out on the sofa. Entropy was increasing; activists sleeping. The decor came courtesy of Volunteers of America, otherwise known as Salvation Army recycling. Only the window plants seemed at all lively.

Mitch Snyder, whose goal is the transformation of society as we know it, came sleepily down the stairs. His heroes are Jesus Christ, Mahatma Gandhi and the Berrigan Brothers. The last he befriended at the Federal Penitentiary in Danbury in the early 1970s. Mitch was charged with transporting a stolen vehicle across state lines, but he maintains that he was merely an innocent passenger in the car. Nonetheless, he was given a three-year sentence and served two and a half. He speaks of his time in prison with rare affection, as though it were one of the few pleasant memories he was blessed with; a time spent reading books, organizing work stoppages, enjoying the inmates' company, fellowshipping, and frequently discussing tactics with the civilly disobedient, some would say publicity-seeking, Jesuit Berrigan brothers.

Snyder was born in 1943 and is divorced, with two teenage sons, neither of whom he has seen for years. His fasts have received high-priority news media attention. He is also the veteran of more frozen nights spent on steamy grates in solidarity with the homeless than he or anyone else can recall. For Snyder the cause may vary from year to year, but what never changes is his passionate conviction that the world he inhabits is basically a rotten one, but one that can be redeemed by concerted social action.

By fasting he persuaded the U.S. Navy to change the name of a submarine, which had been "obscenely" christened *Corpus Christi*. But

236

he failed to persuade the Holy Trinity Church in Georgetown to spend its money on the poor rather than tapestries and organ repairs. The fashionable congregation, perhaps surprisingly, grew tired of being confronted by this gaunt accusatory presence looming over them in their pews. He conducted his fast right there in the church, and you could be sure he would be there of a Sunday morning, nine-tenths martyred and needing only their change of heart to be sent on his way. But they stood fast, dealing Snyder one of his rare defeats. He switched from ideology to pragmatism, went home and had something to eat.

Now he has the homeless as his all-absorbing cause, and he is suing the Federal government in U.S. District Court. We can be sure that *they* won't be rid of him so easily. Unlike the Georgetown parishioners, the Reagan White House is ever-ready to yield. Mitch means business. He gets his media attention from Mike Wallace and "60 Minutes"; his *pro bono* legal assistance from Covington & Burling; his docudrama offers from Hollywood (a $50,000 advance to date); and now he has this huge, dilapidated Federal building full of social outcasts to use as his Diggers, his Albigensians, his gnostic army (his "constituents," as a *Washington Post* reporter put it without irony), in his relentless crusade against Western civilization.

"No-o-o-o-oooo I'm not a Communist," Snyder told me, as though responding for the hundredth time to the same obtuse question—not that I had asked it. He said he was a Christian and by way of corroboration he pointed to a quotation from Pope Paul VI hanging nearby on a wall banner: "Private property does not constitute for anyone an absolute and unconditional right." The Euclid Street building was rented by the community, he said. "We don't believe in owning property. It's not a relationship we wish to reinforce." His voice was low and monotonous, his eyes weary and baleful. His demeanor conveyed patient resignation, as though he were forced to endure time and existence as a heavy burden, a perpetual ordeal, unalleviated by a single moment of pleasure, lightness or humor.

Mike Wallace notwithstanding, Snyder's press often tends to be critical—surprisingly so considering how well attuned he is to the *Zeitgeist*. (But of course he goes too far, even if in the right direction. He is too outspoken, too tactless, too explicit.) He is called arrogant, self-indulgent, a publicity hound. *The Washington Post* accused him of throwing "suicide tantrums." All no doubt true—but beside the point. "Getting your face on television or yourself written about is usually painful," Snyder told me, and I believe him.

Mitch is interesting because he is so fanatical. How many people feel strongly enough about anything to stop eating for fifty-one days?

He ended his most recent (fifty-one day) fast when the Reagan Administration agreed to repair the Federal building the Community for Creative Non-Violence had been permitted to use as a shelter for the homeless since January 1984. A couple of days before the 1984 election President Reagan surrendered. As one writer put it, when Snyder's tactics "became more sophisticated and his demands more concrete, the Administration tried a different response: appeasement."

One person who had been holding Snyder's hand during his fast had been Susan Baker, the wife of Reagan's then chief of staff James Baker. She exerted influence within the Administration. "She's been touched by homelessness and feels need," Snyder said, oddly since her husband is a millionaire. (Perhaps Snyder meant to imply that Mr. Baker is not home often.)

More recently, Mitch and his assistant Carol Fennelly had "tea" with Mrs. Bush and various Cabinet wives at the Vice President's residence. Characteristically, he brought along an urn containing the ashes of a man who had died of exposure on the Washington streets, setting it down amidst the tea cups and cucumber sandwiches. Mitch introduced the remains to the Cabinet wives as "Freddy." Later Barbara Bush drove over to Euclid Street a couple of times with "donations of clothing," and told Mitch that she would like to get more involved in the cause, but, you know how it is, what with the Secret Service detail. . . . I believe that Mitch might have hinted that Mrs. Bush too had been touched by homelessness. . . .

Mitch had by now taken, eaten and digested all the various carrots of appeasement that Reagan's people had handed him, and he wanted more. In fact, he was enraged because, he felt, the Reagan Administration had failed to deliver on its promise to renovate the shelter. So, Snyder was telling me, these were "vicious, cruel people," even "perjurers." They had offered only a ridiculous "patch job," when earlier they had promised a "model shelter." The non-violent community rejected the offer. Next the General Services Administration said the building would have to be vacated (the Hyatt Regency Hotel was offering $11 million for the site). At that point Snyder telephoned Covington & Burling. "They want to do socially relevant work," he said, explaining the ease with which he can attract high-priced legal help. "We're exciting people to work with and we don't bring irrelevant cases. We're always on the cutting edge of things, including the law." He told me that CCNV had at one time or another been represented *pro bono* by all the major Washington law firms, including Wilmer, Cutler & Pickering, and Arnold & Porter.

In July 1985, U.S. District Judge Charles Richey stayed the eviction order and now there is a possibility of a full-scale trial. Over the horizon, of course, is the possibility of a massive increase in Federal welfare programs.

"If we have a model shelter in the nation's capital they know they will have to provide them in all fifty states," Snyder candidly admitted. "And we will encourage that." He pointed to his not-so-secret weapon: "Remember, this is Washington, D.C., and there are 3000 journalists in this town."

I went to the Federal Courthouse for one of the hearings. Mitch came sauntering down the corridors with a small group of his homeless "constituents." A TV reporter asked him where she should station her camera crew. "At what we used to call the Watergate entrance," he replied. The man from Covington & Burling looked like Seymour Hersh with a haircut. A woman was assisting him and hanging around them were interns and activists spending a fashionable summer in the capital. By constrast the team from the Justice Department seemed, well, lacking in compassion.

Mitch told me in the corridor that he had grown up in Flatbush, Brooklyn, and that his parents had been "devout agnostics." His mother was Jewish and his father a nominal Catholic who "despised organized religion." His father had been in the entertainment business, at one point working as a singer with Eddie Cantor. "When he died he was the president of a corporation somewhere," said Mitch, in a tone indicating that he had pretty much exhausted that topic of conversation.

Judge Richey (a Nixon appointee) looked a bit like Earl Warren. He was circumspect, judicious, obviously intelligent, but at one point he wanted us all to know that he was "sensitive to the needs of the people involved—I hope there isn't any question about that." Hearing him discuss the case in terms of need and sensitivity made me wonder if we weren't a society whose "immune system" had broken down in the face of these endless assaults mounted on behalf of the poor, the deprived, the needy, the underprivileged, the minorities . . .

Following the proceedings, Mitch Snyder leaned forward intently in his courtroom seat—the hungry virus probing for soft spots in the body politic.

The jury box was reserved for the news media, appropriately enough. Therein sat Mary McGrory, the left-wing but always interesting columnist. After the hearing she interviewed assistant counsel Florence Roisman outside the courtroom, telling her at one point how thrilling it had been to see her (a woman, no less!) presenting arguments to the judge.

"I can't say how much I admire you," Roisman vouchsafed in return.

On two nights I went to the 185,000 square-foot, four-floor shelter, variously described as a "model of anarchism in action," "more like a turn-of- the-century insane asylum," and a "hell hole," in which its "800 or so residents raise hell." It's not as bad as bedlam, and certainly better than the open air on a freezing night. Currently about 500 people a night stay there—about four-fifths of them men and the great majority black (Washington, D.C. is 75 percent black). By concentrating the homeless in one spot Snyder has succeeded in dramatizing a state of affairs hitherto dispersed. He has put a lens on the homeless, thereby bringing them into focus (a boon for the media).

Some of the men and most of the women represented as "homeless" are mentally abnormal. They are in the shelter as a result of the great emptying out of mental hospitals in recent years—"deinstitutionalization," as it has been called. Snyder is probably correct in saying that most of these people have simply been discarded and left to die on the streets. And the willingness of CCNV members to look after them without remuneration is surely commendable.

Most of the men in the shelter are normal and able-bodied (although some are old), but a good many of these are the indirect victims of a welfare system that works perniciously to dissolve families by depriving men of an unavoidable provider role. With a generous package of welfare benefits available, women with children know that they don't absolutely need men, and in turn the men know that they are dispensable. Thus in a society in which Judeo-Christian values are weak, the (ubiquitous) forces of familial dissolution are liable to prevail over the (now optional) ties of matrimony. The result is demoralized males who take to the bottle and the street. The problem is severe with blacks because welfare has been aimed particularly at them, and blacks have not yet found the leadership necessary to escape this destructive aid that comes in the guise of compassion.

Mitch Snyder gives a grudging assent to this anti-welfare analysis, but he is not happy with it because it singles out the welfare state for criticism when in his opinion "nothing works" and therefore the entire society has to be transformed. Nonetheless, Mitch is correct in perceiving that the welfare bureaucrats have no interest in doing anything about these demoralized males who prowl the streets. They just don't want to be bothered with them because they are difficult and intractable. Government compassion is strictly a nine-to-five affair. Welfare employees in Washington in their tens of thousands have things so arranged that, like university faculty without students, they only have to deal with files, memos, charts, budget numbers and meeting-boredom—not actual people at all. And certainly not this strange male flock chattering away with incongruous bravado at midnight outside the shelter entrance.

Larry, 35, a Washington, D.C., native, told me that he had stayed at the shelter since April. He said he was "from the middle class"—his father was a baker with Giant Food. Larry graduated from high school, was well spoken, sober, and seemingly responsible. "I work pretty regularly, but my pay is strictly commission," he said. "So I haven't been able to afford to put a roof over my head." He works for a street vendor on K Street, N.W. Now the mayor has raised the license fee for vendors from $15 to $606, Larry said, and this has duly driven many of them out of business. (Remember the Laffer Curve?) Some of these ex-vendors are now staying in the shelter. (The irony is that Larry is black, as is the mayor, who responded in part to media huffing and puffing that too many of these vendors are Asian, not enough of them black.)

Larry said that he has a seven-year-old son but doesn't know where he, or his former wife, now live. Nor did he know whether his wife received welfare. "I met her," he said, "fell in love, got her pregnant, and married her, but it just didn't work out. Wasn't meant to. With a whole lot of men relationships don't work out—just fall apart." It was "fifty percent my fault," he said, and clearly hadn't considered apportioning any blame to the state.

That day some of his "measly possessions" had been stolen from under his bunk, which is in a tumble-down room with no door and eleven other bunks. "Security is definitely a problem," he said. "It could be lessened if the dorms were smaller." Nevertheless, he was a strong supporter of the shelter and hoped for its renovation. "It gives a person who has taken a fall a place to bunk and get himself back together," he said. He has known about 35 people to do so while he has been at the shelter, and he plans to be gone himself soon. "We're all God's children, and we all need a little help from time to time."

* * *

Mitch showed me around the shelter and told me that the 35 CCNV people who work there are unpaid. "No one should work for money," he said, adding that the community had to date spent about $200,000 in donations. (His Hollywood advance is likewise going toward shelter bills, another CCNV member told me.) The same services provided by government would have cost at least $10 million, Mitch said. "We threaten the charity providers." (Usually he calls them "poverty pimps.")

As we walked around the corridor maze several men approached Snyder and either thanked him or asked his advice on qualifying for a library card or getting Food Stamps.

"The Reagan Adminstration professes opposition to the centralization of power," he said at one point. "We're trying to get them to live up to their principles. We have to preach conservatism to the Reagans."

Snyder's point is that if we all looked after one another—following the example set by CCNV—then we could dramatically shrink government and save "not millions but billions of dollars." Nevertheless, as he sees it, the government does have a responsibility—to provide buildings.

"Once the building is fixed up we have absolute responsibility for everything else. This is our shelter and we're running it. We pay the bills, including the electricity bills, and we're contributing our services. Government officials are free to come in and look at it because it's their building, but they are not free to run our programs because if they had the answers, we wouldn't be providing the solutions. We know the nature of the beast we are dealing with. That's why we are not tax-exempt. We are not going to be beholden to the U.S. government, because it's a ridiculous structure: It's big, it's burdensome, and it's inefficient. It takes the place of citizens fulfilling their legitimate responsibilities, and I don't like it for all of those reasons."

I reminded Snyder that there was an obvious Trojan Horse problem. The government could yield today, providing buildings in all fifty states as he wished. But after some time had passed they could easily return to their buildings, thrusting aside all volunteer activity and embark on a massive expansion of welfare programs. Snyder agreed this was a danger, but his attitude seemed to be that he wasn't going to think about it. The government's role was to provide buildings, a "one-time deal, a capital expenditure," he repeated, and anyone who said otherwise was "either immoral or an anarchist."

A strange-looking fellow came stalking down the hall with his clothes on back to front and his legs spread-eagled as he walked.

"Before this place opened, he'd be on the streets," said Snyder. "Great
country."

Many walls, either hardboard or cardboard, were smashed and on one
was written "Put Koran In Public Schools." Twice I overheard disparaging conversations about Jews. ("Overheard" is perhaps misleading
because they weren't exactly whispered.) Men and women occupy separate quarters and Snyder insists that they must remain separate in any
model blueprint. This seems anomalous at first sight; a rejection of the
communitarian world view. In fact it shows Snyder is serious—willing to
put reality above ideology when it comes to practical detail.

Mitch showed me his room in the top-floor staff quarters: a rather
monkish cell with a single bunk, an alarm clock and a cat. He told me that
he was "a Leo," and that he "preferred cats to rats." Communitarianism is
either orgiastic or ascetic; Mitch clearly belongs in the latter camp. He
takes a dim view of casual sex. On abortion he speaks like a politician ("I
worship life but I also don't like limiting people's choices"). He doesn't
smoke, drink alcohol or coffee ("cash crops tend to be dangerous") and he
worries that he ingests "more protein than I am entitled to." He sub-
scribes to an "era of limits" world view in which he uncharacteristically
defers to "experts" who have decreed that the world cannot support
much more than its current population.

I told Mitch that of course his collectivist ideas have been put into
practice rather widely in the 20th century, notably in the Soviet Union.

He was ready for that one. "They're political, we're religious," he said
quickly. "The political approach results in broken, twisted models." Of
course Snyder's own approach is highly political, making full use of such
institutions as the courts and the media; just as CCNV's very effective
food scavenging operations depend (whether Mitch knows it or not) on
the existence of a market economy.

"The answer is only found in communities which acknowledge the
oneness of one another, and that will only come about when the distance
between human beings is reduced to nothingness . . ." and here Mitch
plunged into his warm Utopian pond. "And when that happens," he said,
"the world will *instantly* become a better world. So the truth is that living
in community is the only way we can make it."

Mitch claims that his approach is religious, but like his father he
despises "organized religion" (churches are nothing more than "corpora-
tions," he says) and his detailed plans for the model shelter omit religion
completely. He laid out for me the "full range of services" that his friends
and allies have already offered to provide free of charge: "Medical,

dental, mental health, job counseling, drug and alcohol rehabilitation, a non-profit employment agency, benefits counseling [how to get welfare], and ongoing adult education, which means literacy training." It sounds a lot like the old Department of Health, Education and Welfare all over again. No mention of religion, and if he were to try to include it you can be sure the ACLU and People for the American Way would have Snyder in court before you could say "wall of separation."

"How you doing?" one of the homeless asked Mitch.

"Tired," he said.

"Don't you ever rest?"

"I'll rest when I'm dead," he said. "That will go on for a very long time."

I asked if he believed in an afterlife.

"I don't think in those terms," he said. "I believe in hell. I believe we're there right now. This is it. It couldn't possibly get any worse than this? Look around you. Look at where you are! What could be worse? To be dead?"

Did he mean life in general, or the shelter in particular?

"No, the world," he said. "This earth. The limitations of the flesh. The struggle. Everybody struggles with the same infinitely stupid questions, because we're imperfect creatures. Because we're of the flesh. So we pay a price for that. But I don't put it in terms of an afterlife. There's a connection between everything that was, is, and will be. The bottom line is that all human beings are members of the same family, the human family is part of whole creation, and we have to exist in harmony with it and we're not doing that because we're out of sync, because we're living unnaturally, and what's most unnatural is that we're not living collectively."

With that gnostic message he accompanied me to the door. "I want to find the kingdom," he said. "To bring the kingdom into existence. There's nothing to move on to. [I had raised the possibility.] It's with people like these, people I have lived with and worked with for ten years that I will find my salvation, and it's in the political arena that I will transform what I have seen and experienced into something that will touch other lives."

At the door the topic returned to his Holy Trinity hunger strike. He said that "a lot of powerful people in Washington have never forgiven us for that. And never will. That's good—they shouldn't. They were right. What we were doing there was very dangerous. And what we're doing here is very dangerous."

King David's Royal Family

Ralph Dyer drove me to the four-story house just off Christopher Street in Greenwich Village, where Mother Teresa's Missionaries of Charity run a home for AIDS patients. The house is an archdiocesan building, formerly the rectory of St. Veronica's Church. Outside was a sign, "Gift of God, Missionaries of Charity." It was opened on Christmas Eve, 1985.

We were admitted by a volunteer worker. To our right as we entered was the chapel, dominated by a large, realistic crucifix on the wall, next to the letters, I THIRST. In the hall was a picture of Mother Teresa above the caption: "Unless life is lived for others it is not worthwhile." Posted nearby was a sheet of paper listing names and "departure" dates of the nine men who have died since the house was opened. The last date was July 2, 1986, and beneath it was ample empty space. "Heaven Is Our True Home," it said above the list. "Welcome Home My Sons."

Two volunteers were working in the kitchen in back. Otherwise the house was silent. Fourteen men with AIDS were living there. The hallway and staircase were decorated with Easter-style streamers in yellow and white, celebrating the fact that one of the men was going to be baptized that afternoon.

Four or five nuns (or sisters as they are usually called) work in the house, devoting their lives to the patients. Sister Dolores, the superior of the house, came out to greet us wearing the white, blue-trimmed sari of the Missionaries of Charity. I also briefly met a sister from Portsmouth, England, and another from Chicago. But they remained industriously about their business, and didn't encourage small talk or personal questions of any kind.

Sister Dolores, who is from India, ushered me into the small front parlor. She told me that she entered the novitiate of the order in Calcutta in 1960, but beyond that she disallowed questions about herself.

245

One point that she wanted to emphasise at the outset was that they don't try to convert anyone who doesn't specifically ask for it. Evidently the Missionaries of Charity have an ecumenical view of salvation.

"It's according to whether they say 'I want,'" said Sister Dolores. They have Mass and say the rosary every day, and I gathered that most of the patients do attend. There is also periodic religious instruction. But everything is voluntary. She said that most of the men have died in one or another of the nearby hospitals, because the sisters don't have much in the way of sophisticated medical equipment. "But practically all of them wish to die at home," she said, "because they know we will keep them company."

Here are abbreviated versions of written accounts kept by Sister Dolores, recording the histories of three men, all white, who died recently in the sisters' care.

Dennis: Aged 42, at house for five months. Had a long history of drugs. Expressed a desire to die in the house and was the second to do so. More and more was coming to God. When already very weak was unexpectedly at Mass one day. Was visited both by Cardinal O'Connor and Mother Teresa. When he had a headache he said he thought of it as the Crown of Thorns. The day before he died he received the Sacraments and Father Greg gave him the Last Sacraments that afternoon. Sister Dolores read the Gospel as he expired in great peace. His face reminded everyone of Jesus on the Cross.

Harvey: The first to die in the house. 38 years old. Long history of drugs. Military service in Vietnam. From the beginning was grateful to be with the Sisters. "Call me Brother Harvey," he said. Taken to hospital one night very sick. "I want to be baptized as a Catholic." Attached great importance to the rite. He was given white shirt and trousers to wear. Wore rosary around his neck. "Teach me to pray," he said. When asked if he wanted to go to heaven he said, "Yes, but I don't want to leave this house." But it would be better in heaven. "If you say so, I'll go." He expired most peacefully. Most of the men said they wanted to die that way.

David: Died in St. Vincent's Hospital, aged 34. Came to house in March, 1986. Started drugs at age 14, prison record. Presbyterian background. At first he was so demanding that the sisters called him King David, which pleased him. Would go out to get more drugs. One day he prayed the Our Father in bed, said it was the first time he had ever prayed. Taken to hospital and received First Communion a few days later. Visited by Mother Teresa and Cardinal O'Connor, who asked David to pray for him. Once he said he would take all our sins, put them together like clay, and put it before Jesus on the Cross. "I am in close contact with God." Died early the next morning. His mother said later that David had taught her much, and she wanted to be baptized as well.

Sister Dolores showed me where the men slept—two to a room. In one room a pale, emaciated man with spinal meningitis looked across at me in an amnesic fugue, groping at his goatee as though trying to recall something long lost.

In her comments, Sister Dolores was marvelously appreciative of the individuality of each patient, as though she not only had King David in her care but his royal family as well. Several times she repeated that they were a family.

She took me to a room where eight men were seated, most of them wearing dressing gowns or T-shirts. All were black, Puerto Rican or Cuban. I sat next to Tom Ryan, an intuitive soul and no doubt a dude in his day. He was the one who would be baptized that afternoon. Like Harvey, he wore a rosary around his neck. A 40-year-old black man, he had been in the house for three months (since April) and had been diagnosed with AIDS in January. He was placidly flipping through an old *National Geographic*. Next to him sat his good friend Juan, slightly built, wearing a Nike T-shirt. Juan was lost in his thoughts.

July 22, two days away, was the birthday of a black man called Walter, who had been in the house only two weeks. "Does anyone know whose feast day that is?" Sister Dolores asked. "It begins with an M."

But curmudgeonly Walter refused to be flattered or pleased by this attention to his birthday. He didn't want any fuss or bother. "It ain't nothin' to celebrate, sister."

Later, Tom Ryan told me that Walter was "searching, inquisitive." And he was changing. "When he first came he couldn't sit down for five minutes."

"Mary Magdalene," said Sister Dolores, answering her own question. "Does anyone know who she was?"

"She hung out," said Tom, almost to himself, as he looked down at a pride of *Geographic* lions in his lap. "Loose woman."

Sister explained that Mary Magdalene had been the first to see Jesus after the Resurrection. Walter listened attentively as she told the story.

That afternoon Tom told me about Juan. "My ace man, a very blessed man I think. He was El Diablo. Hot blood. Used to wear a pony tail. We shared a room together. One day he was lying there and he said, 'They used to call me Indio.' I said, 'I remember you, you was a killer, man!' The terror of South Bronx. One day he wanted to go back—" and here Tom explained that "all of us have tried it, and we all have our scars to show for it." Ignoring the sisters' advice, Juan went to the South Bronx by subway.

"On Westchester Avenue he fell on the street," Tom went on. "He said, 'Tom, I got tired. Didn't do much walking.' His face was bleeding where he fell. No one in the crowd wanted to help. Made me think of Christ on the way to Calvary. Two white people got out of a car and put him in an ambulance. He used to give the sisters hell at first. 'Don't tell me what to do!' Now he's as gentle as a lamb. But with so much fire—fire for goodness."

Ralph Dyer drove us across the Brooklyn Bridge to pick up Tom's wife and two children, who were to attend his baptism. With us was Anne Petrie, who with her sister Jeanette recently produced the feature-length film "Mother Teresa," five years in the making. It is due to be shown on television and released worldwide on December 8th.

Tom had used drugs in Vietnam "to ease the strain," he said. On his return to the United States there were "baby killer" signs in Oakland, rotten eggs and vegetables thrown at soldiers. "Men were changing clothes in the plane so as not to be seen in uniform." He continued using drugs (heroin and cocaine), "all for a fleeting feeling, and then you go out and get more money to chase that feeling. I've put a million dollars in this one arm over 18 years. Life got to be a big dull pain."

Tom grew up in Brooklyn. As we drove along he pointed out "the famous Brooklyn library," and "the famous Brooklyn Museum." His mood was one of detachment rather than mockery. He said that he went with his parents to the Church of God in Christ in Harlem. Sometimes he would spend all day there. "But the people weren't living what they were talking about," he said. "Listen to the men talk beautiful and see the people do the opposite." By contrast he noted the "sympathetic influence" of Mother Teresa's sisters. "They live their religion. And I want to be a part of that. I can't do no great works. And yet through faith my understanding slowly opens up. I guess I've been starving for it so long."

He had been fortunate enough to meet Mother Teresa at the house. She was "like a quiet storm that will shake you," he said. "She says little things—but from them come oak trees."

We drove through the trampled lots, shattered glass and derelict structures of Brownsville—many of them boarded up with plywood. Tom said how much the neighborhood had deteriorated. Eventually we came to a project high-rise, where his wife and two children were waiting for us, all dressed in white for church.

Most of the men I had seen that morning were waiting in the chapel. Father Greg turned out to be a cheerfully extroverted Australian, who in the course of his sermon actually used the much caricatured expression "fair dinkum," (meaning, roughly, "true"). Like the sisters, and following the Indian custom, he was barefoot in the chapel. As he proceeded he explained every procedure simply and yet seriously.

"Do you renounce Satan and all his works?"

"I do."

Tom had become Thomas and he had chosen the additional name of Peter.

". . . I baptize you in the name of the Father, and of the Son, and of the Holy Spirit. . . ."

Thomas Peter stood in a white plastic receptacle as the water was poured over his head—enough of it to streak his dark-colored work clothes. Then he went outside and returned, like Harvey before him, wearing white trousers and shirt. His wife, daughter and especially his son hugged him emotionally.

It was hard to tell what impression it all made on the other patients, most of whom sat impassively throughout the ceremony. Juan received Holy Communion.

Earlier I had talked to Tom alone in the chapel. He sat and talked with a relaxed candor and tranquillity for an hour or more. He said he was now completely free of drug addiction. He had received "hundreds" of blood transfusions while on kidney dialysis, he said. So maybe he got AIDS that way. "But with an I.V. drug record, they blame it on the drugs," he said.

"I blame nothing. I don't think I would have ever come to God without this. They would have found me in the street. I used not to be able to talk about these things. There are people who want what I am receiving. But it took all these years of pain and suffering to find it. I can get on my knees and pray and feel refreshed. No better feeling in this world. And inside I want to tell the world."

He said that some resolutely exclude the truth, "keeping up a front even to the grave. They do. They know what you are saying is true. But they despise it. I used to. Broke all Ten Commandments many times over."

To others with the same illness he would say: "Now is the time to give back something to God, even if only prayer. I imagine that each and every one needs reassurance to try to grasp Christ. But don't despair. Just think on it. Because He is the only way out of this. And any miracle that comes will come through him. Even in sickness you can bear the pain, because you share that pain with Christ. I've had so many people before me who died. Some had longer tails than the devil. In the end they ask for Christ. They say 'Oh God have mercy!'

"Strange sounds from men who cursed and blasphemed. But in the end they grasped the peace."

Safe Sex at Stanford

The great condom hullaballoo suggests that AIDS is regarded as a greater threat to the sexual revolution than to the health of America. If the spread of the disease is indeed as alarming as everyone claims, why doesn't *someone* in the mainstream media urge that we practice sexual abstinence before marriage, and monogamy within it? But such an appeal would tacitly endorse Christian morality. It would give aid and comfort to Jerry Falwell and Phyllis Schlafly. Nonetheless, we're being told daily that we have on our hands a crisis of epidemic proportions. So why not fight the disease with the most effective barrier to its transmission—abstinence? We don't have to consent to Christian morality in order to take advantage of its utility.

But we can't have that, can we? It would be terribly unprogressive to give up all those hard-won '60s "gains" against conventional morality. Better to march forward with the sexual revolution and reduce its risks with condoms than to take the retrograde step of making even a pragmatic case for Christian morality. . . .

Such were my thoughts on February 16, 1987, when I opened *The Stanford Daily* and read the front page headline: "Campus Launching Condom Week."

> As part of National Condom Week at Stanford, the AIDS Education Project will distribute 500 packets of condoms today and tomorrow in White Plaza [centrally located on campus] between 11 a.m. and 1:15 p.m. Each packet contains eight brands of condoms, among them 'exotic pink and black condoms from Japan, smooth and ribbed condoms, and a condom with Tahiti colors,' according to senior Meg Richman. Students will be asked to rate the condoms according to color, size, shape, texture and taste.

251

The rating forms are due to be returned Friday, and those interested could know by next Monday which condoms Stanford students like best, according to Richman.

That's mouth-taste, not good/bad taste.

"Stanford students appear caught up in the excitement," noted Baie Netzer, *The Stanford Daily* staff writer.

White Plaza is only a short distance from the Hoover Institution so I went to take a look. The first thing I noticed was a table set up with a big sign reading: STANFORD SAFE SEX KITS.

Safe? Only a couple of days earlier I had read in *The New York Times* that "condoms are not foolproof. . . . In one study two cases were documented in which the virus passed between partners who used condoms over an extended period." Condoms often don't prevent pregnancy and women are fertile only once a month. The AIDS virus is potent every day.

You would have thought with the "torts," or liability explosion in our courts, Stanford would be more cautious about lending its name to the distribution of "safe sex kits" on campus. What if a student came down with AIDS after using one? Might he not be in a position to sue?

I was handed a safe sex kit. It contained seven condoms and a mauve brochure reading:

YES, WE ARE SERIOUS

"The purpose of this first annual condom extravaganza is to provide an opportunity for you to sample the many different types of condoms currently available. We want you to have safe hot sex that is also healthy sex. Given the risks out there (unwanted pregnancy and incurable sexually transmitted diseases) we want to help you develop an erotic sex life that at the same time displays responsible, safe and healthy behavior between you and your partners."

"*The Great Condom Rating Contest* is brought to you by the Condom Promotion Campaign of the Stanford AIDS Education Project—P.O. Box 8265, Stanford CA 94305."

I spoke to Ken Ruebush, 22, a senior and German Studies major who is coordinator of the Stanford AIDS Education Project.

"I see education as the only means of preventing this tragedy," he said. "We don't have a medical solution."

How about abstinence?

"I think that people are basically sexual entities," he said. "I don't think it makes sense and I don't think people will follow that advice." (They certainly won't if society no longer has the courage to give it.)

"Abstinence is 100 percent effective but not 100 percent realistic," said Meg Richman, a senior from Los Angeles, who was also handing out kits.

Are they not promoting promiscuity in the guise of "safe sex"?

"Give him the shoe analogy," said Daniel Bau, a junior. "Giving out condoms promotes sex is like saying giving out shoes promotes walking."

Ruebush said he had been accused of giving the equivalent of sterile needles to addicts. "If we had a big drug problem on campus," he said, "I would be in favor of giving sterile needles to addicts."

Notice, incidentally, that in the case of drug addiction the conventional wisdom, formerly "controlled use," has now been toughened up: "Just Say No!" Similarly, millions of anonymous alcoholics have not given up on the freedom of the will. Nor have addicted cigarette smokers who try to quit. Only when it comes to sex is the exercise of willpower impugned as an inappropriate application of moral principle to bodily function.

As for Acquired Addiction Syndrome—the idea that abandoning oneself to sexual temptation only increases one's addiction to it—this belongs in the growing category of Thoughts that are Forbidden in Our Day. Nonetheless, it tends to be particularly true of unnatural forms of sex and helps to explain why so many homosexuals have so many partners. They have yielded to temptation so many times that the fires of lust burn within them. In the end they do indeed find it difficult to control themselves. They are sex-addicts, and some (those, for example, who go to "Courage" meetings, a Catholic organization analogous to AA), wish they weren't.

"Sex is not a medical problem to begin with," Ken Ruebush told me. (Nor was smoking. Nor was cocaine.) "People have to make their own decisions whether sex is morally wrong for them." (It would be interesting to make a list of those topics at Stanford that are denied the standing of moral relativity—apartheid, for example. And increasingly—smoking.)

Ruebush said that the Great Condom Rating Contest was really "just a gimmick to get the word out—a way of de-stigmatizing condoms."

What we are seeing is the subtle use of ridicule to disarm shame. At U.C. Berkeley there were "five events," including a "Water-filled Condom Throw." (Incidentally, who organized this simultaneous, nationwide campaign—"seven days of serious educational efforts and festive foolery" at "scores of colleges," as *The New York Times* put it, without identifying the organizer?)

Even before AIDS was known, Eric Hoffer told me in San Francisco that what worried him was the breakdown of shame among homosexuals. It would end in disaster, he said. "There cannot be civilized living without shame." Now we are in the midst of a further assault on the natural barriers of modesty, in the guise of "festive foolery."

"I've had friends who died of AIDS," Ruebush told me in White Plaza. I asked him how many people in the Stanford community had the disease.

"At least 20," he said. "And that is a figure I heard nine months ago."

According to *The Stanford Daily* of October 6, 1986, "between seven and ten cases of AIDS in Stanford students have been reported" to the dean's office. I was informed by three people that two doctors at the Stanford Medical School have died of AIDS. Bob Beyers, director of the Stanford News Service, reported in October, 1985, that at least ten members of the Stanford community (including three faculty members) had at that time been diagnosed with AIDS. He told me that "it would not be possible to give you exact figures," because typically what happens is that students are likely to "withdraw from the community and die elsewhere."

Ruebush said the same thing—students with AIDS go away to die. (Like wounded animals—the image comes unbidden.) He suggested that I talk to Dr. Paul Walters, for four years the director of the Cowell Student Health Center.

"You're going to have trouble getting those numbers," he said, when I asked him for the AIDS figures.

Any particular reason?

"Well, the number is small enough that people are concerned about the confidentiality."

"Ken Ruebush said at least 20."

"Did he."

"You neither confirm nor deny?"

"I wouldn't."

"Wouldn't you say it's of some interest for the Student Health Center to know how much AIDS there is or has been on campus?"

"That's not a question to ask me."

"Obviously it *is* of interest."

"Whether we're going to discuss it is another matter," he said, chuckling lightly, as though confident of victory in this particular fencing match. "This is a highly sensitive matter."

"I was told two Stanford doctors died of AIDS."

Dr. Walters would neither confirm nor deny.

Meg Richman told me that the AIDS Education Project is an out-growth of the Gay and Lesbian Alliance at Stanford (GLAS). "It was extraordinarily useful that we were able to work with GLAS in educating the community," Dean of Student Affairs James Lyons said a few months ago. "Stanford was very fortunate in having a well-organized gay and lesbian community when the [AIDS] phenomenon hit us."

The role that the Gay and Lesbian Alliance has played on campus may be gauged by the invitation they extended in March 1984 to a once-imprisoned Dutch parliamentarian and pederast named Edward Bron-gersma to speak on campus. He told an audience of fifty that sexual relations between adults and children should be permitted by law, "pro-vided the child consents." For example, he said, sex between a 30-year-old man and a seven-year-old boy would be fine. "If we see sex and lust as a good thing," he told the Stanford audience, "I don't see why we should separate it as a category apart from other things."

Only a few weeks later, university president Donald Kennedy spoke at ceremonies for the 10th Annual Gay and Lesbian Awareness Week. "I'm very glad to be here," he said. He wanted the company assembled to know that he was president of the *whole* university, not just part of it. "I'm grateful for what a lot of the people in this room have done for this university," he continued. "In a general way, you've provided all kinds of services and intellectual enlightenment to this place."

Does pedophilia fall into the category of intellectual enlightenment? For all I know, it does. I do know that if a racist were invited to speak on campus the group that invited him would be in trouble with the Univer-sity administration.

In May, 1986, according to an engraved invitation, "The Gay and Lesbian Faculty, Staff, and Provost James N. Rosse, request the honor of your presence at a reception celebrating the sixteenth anniversary of the Gay and Lesbian Alliance at Stanford," the reception to be held at Avocado Grove, Humanities and Sciences, Inner Quad Building I.

"For information call Prof. Ron Rebholz," it said on the invitation, so I called him—nine months late, to be sure. Rebholz is a member of the English Department and of recent years a frequent critic of the Hoover Institution, which he thinks is "too political" and too much in favor of President Reagan. He said it would be okay for me to come and talk to him. Rebholz, incidentally, is the university's resident Shakespeare expert. He told me that his leftist politics stemmed from his father's frustrated union organizing activities while working for many years in a menial capacity for a bank.

I asked him about Stanford's "Gay and Lesbian Faculty." How many members did it include?

"About fifteen," he said, adding that the university administration had been "very supportive of gay people and AIDS victims." But he said that he didn't think that sexual behavior "should be the object of anyone else's interest." But why not if it transmits a fatal disease?

I asked him if the university had any particular rules about faculty members having sexual relations with students.

"They have eliminated 'moral turpitude' as a standard for firing people or for getting tenure," he said. "The unwritten code, generally, is not to be involved with students whom one might be grading."

Leaving Rebholz's office, I walked across the main quad and examined a piece of sculpture on display near the Mathematics Department, "Gay Liberation," by George Segal. The sculpture was installed in February 1984 on loan from the Mildred Andrews Fund, and was awaiting transfer to the Harvey Milk Plaza in San Francisco. It consists of two life-size couples—two men standing close, one man's hand around the other's shoulder, and two women seated on a park bench nearby, one woman's hand on the other's knee. The bronze (white patinated) realism of this work stands in stark contrast to the familiar array of cretin crankshafts elsewhere littering the campus. Apparently, when the cause is progressive, realism is appropriate.

Two weeks after it was installed, a "lone vandal" came by one night and hammered at all four statues, but did little damage. "We are outraged," said the "press spokesperson" for the Stanford Gay and Lesbian Alliance. "The four figures depicted in this sculpture are, in a very real sense to us, our public representatives. The fact that they have been violently abused is a symptom of the homophobia which pervades our society, where harassment . . ." etcetera.

Stanford Art Prof. Albert Elsen said: "The deed speaks for itself."

At the time of the condom distribution, James Rosse was acting president of the university (with Donald Kennedy on sabbatical). About a month earlier the local papers had reported the death of the dean of the university's Earth Sciences Department, Allan Cox. Rosse was said to be "very surprised" when he learned that this death was being investigated by the San Mateo County Sheriff's Department as a possible suicide. An avid cyclist, Cox had ridden his bicycle into a redwood tree in the hills above the campus, and he had died immediately. A few days before this, Cox had been told of a pending criminal complaint of child molestation against him.

The Stanford Daily reported:

"The father of the alleged molestation victim studied for his doctorate under Cox at Stanford, according to a report yesterday in the *San Francisco Chronicle*. The parents reportedly said Cox had sexual contact with their 14-year-old son when the family was living on the Peninsula, and claimed that Cox had been sexually involved with the now 19-year-old since then."

None of this came out in *The New York Times* obituary of Cox, nor in the lengthy, friendly *Los Angeles Times* article on Stanford, heralding the university's $1.1 billion centennial fund-raising drive.

Perhaps acting president Rosse would have been less surprised by these developments if he had attended the Gay and Lesbian Alliance-sponsored talk given by the pedophile Dutchman. Was Cox present on that occasion?

All of this would seem to raise serious questions about the responsibility of the current Stanford University administration. Just imagine the uproar if anything comparable could be imputed to Hoover Institution Director W. Glenn Campbell, whose every statement is subjected to obsessive scrutiny for signs of conservatism. On the day when Stanford students were rating condoms for "taste," *The Stanford Daily* reported as a front-page story:

> The Faculty Senate will consider tomorrow a motion to censure Hoover
> Institution Director W. Glenn Campbell for claiming that the location
> of the Reagan library on campus allows the University to "boast" of its
> "Reagan connection."

The censure motion passed unanimously. A few months later the faculty forced the Reagan Library out of Stanford.

Sidney Hook at Hoover

The phrase "sweet reason" took on new meaning for me when I became acquainted with Sidney Hook at the Hoover Institution. He rejoices in reasoned discourse, and will engage you with a kindly open-mindedness that is rare among intellectuals, not to mention intellectuals in their 80s. Sidney will be 85 in December, 1987. He once told me that he hoped I wouldn't be disappointed to learn that he was a "naturalist," which meant the same thing as materialist, philosophically speaking; but he avoided the latter term because it also confusingly suggests the hedonistic consumer of luxury goods. Nothing could be more misleading in Hook's case. I was touched to find him solicitous of the faith of others, suggesting when we started to discuss the problem of evil that I might want to consult a theologian rather than be exposed to the perils of doubt.

But I told him that no one should fear reason. He replied that St. Thomas Aquinas would certainly have agreed with that. "I have always been partial to the golden reasoning of Aquinas and his inspired common sense derived from Aristotle," Sidney Hook writes in his just-published autobiography, *Out of Step: An Unquiet Life in the 20th Century.* "I find myself intellectually more comfortable in disagreeing with the clearly enunciated propositions of Catholic Thomists than puzzling over the fuzzy confusions of some modernist Catholics, who have been infected with Protestant and pantheist heresies and lack the courage to avow it."

In recent months I have had the opportunity to discuss these age-old philosophical problems with Sidney Hook, the 1985 winner of the Presidential Medal of Freedom. He taught philosophy for many years at New York University, where he was chairman of the Philosophy Department. He must be one of the very few secular humanists both clear-headed enough to know that that is what he is and courageous enough not to reject the label. (But he does deny that secular humanism is a religion.)

About once a week I would hear Sidney out in the corridor buttonholing Hoover fellows and staff, to take issue with the latter-day Catholic bishops. "They act as though they want to promote socialism," he told me one day. "I wish they were more interested in religion!" In his autobiography he describes a Dialogue Between Catholics and Humanists at a Catholic retreat in New York City. Evidently then still thinking of the Catholic Church as though personified by Father Charles Coughlin, he was "shocked to discover a change in the quality of thought and behavior of the Jesuits." One Jesuit defended the Chinese Communists, then still worshipping Mao. "To be confronted by a Jesuit priest who criticised me for protesting the ruthless persecution of Catholics and other missionaries on the ground that the Communists were trying to help the exploited Chinese was an extraordinary experience. For a moment it reduced me to a stuttering inarticulateness; then I let fly at him. Although I did not say it, I felt he was a disgrace to his cloth, which was a strange emotion for an unbeliever."

I was eager to explore further Sidney Hook's beliefs, or lack of them, and so had several discussions with him. One evening he was wearing his navy blue beret and was carefully buttoning up his raincoat as he prepared to walk home. A slight, almost a frail figure, he will frequently clamp his eyes shut and grimace as though in pain while making an argument. He picked up his briefcase and we set off for the fifteen-minute walk across the Stanford campus.

"At the age of 12 I discovered the problem of evil," he told me. "How can you reconcile God's goodness and infinite power with the suffering of the innocent and the success of the infamous?"

He was about to be "Bar Mitzvah'd," or confirmed, when he put up some resistance because of this "bone of evil which has been choking theologians through the ages." Eventually he surrendered to his parents' argument that he would disgrace them in the eyes of the Jewish community, and went through with the ceremony.

His father, Isaac Hook, came to America in 1882, an immigrant from Austria near the Bohemian border. The family settled in Brooklyn, where Isaac worked in the garment industry dawn to dusk (but with protracted layoffs). He taught himself to read and write in two languages, but never went to school. He observed the Sabbath day (although Sidney noticed that he sometimes smoked when he was not supposed to) and he prayed in the morning.

"How do you know there is a God?" the future philosopher and friend and correspondent of Bertrand Russell and Albert Einstein would ask his father.

Isaac would reply, "How do you explain the existence of the world if it was not made by God?"

But young Sidney would reply: Who made God?

"You can't ask that question," Isaac would say. In which case, Sidney would riposte, his father couldn't ask who made the world. It seems that Sidney would get the last word.

"Who am I to tell you any different?" his father would say.

"If he had had my opportunities . . ." Sidney mused, a cold wind blowing as we walked along.

Take a suffering child, Sidney said, returning to the problem of suffering. Why would a good God punish that child? Because of evil done by his grandfather? And of the Biblical promise that ancestral sins are visited on later generations, Sidney said, his voice rising with a touch of indignation: "What kind of justice is that?"

Sidney told me that his mother had lost her first-born child when the child was two years old—something he doesn't mention in his autobiography.* (It is a rewarding work, but it contains nothing about his family background.) His mother had bought a puppy for the child to play with. But the puppy upset a pot on the stove, scalding the child to death.

Occasionally, his mother would ask. "How could God have permitted it?" But "she didn't abandon her religious practice," Sidney said. "She remained a conforming member of the Jewish community." But he speculates that it shook her faith. He said that he learned of the tragedy when he was about five years old.

As we walked past the Eucalyptus trees near the Green Library, I mentioned the episode narrated in St. John's Gospel, when Jesus saw a man "blind from his birth." The disciples asked: "Master, who did sin, this man or his parents, that he was born blind?" Jesus answered: "Neither hath this man sinned nor his parents: but that the works of God should be made manifest in him."

"Why would blindness make the works of God manifest rather than

---

*But it is mentioned in *The Courage of Conviction*, (Dodd, Mead, 1985), edited with an introduction by Phillip L. Berman, in which "thirty two of the world's best-known men and women intimately and honestly address the questions, What do I believe? and How have I put my beliefs into practice in daily life?" In addition to Sidney Hook's, the book includes interesting contributions by Edward Teller, Edward O. Wilson, Benjamin Spock, Mario Cuomo, Jane Goodall and others.

extraordinary vision?" Sidney countered, again with a touch of indignation in his voice. "Think of the suffering of those close to him, his family
and his friends. 'The heavens proclaim the handiwork of the Lord.' All
right, I can accept that. But gratuitous suffering? Some suffering can't be
accepted."

I said that he was insisting that the ways of God be scrutable to man;
that the "books" of justice be balanced with each succeeding entry; that
every step, every event in history be balanced in the scales of justice.

"Not every step," Sidney said. "But on the whole. Of course it's
unreasonable to expect justice at once. I don't expect the innocent man to
be vindicated tomorrow. But I do expect him to be vindicated before he
goes to the scaffold. I don't expect the tyrant to be struck by a thunder-
bolt. But I would expect, if there is a God, that he would not die
comfortably in his bed."

Small children cannot understand why things happen the way they do,
I said. And their parents know it is impossible to explain everything to
them. At some point a parent just has to say to the demanding child,
"Because I say so!" Perhaps that is our relationship with God. Is it not a
form of pride to insist that our understanding be on the divine level?

"And my answer to that is: Is it expecting too much to ask that at some
point as adults we would be blessed with the understanding to see the
justice, if it exists, in the succession of events over which he rules so that
not a sparrow falls to the ground without the Father knowing?"

But then, as we walked on further, he pushed the topic aside and said that
the struggle with totalitarianism in the Soviet Union was a more urgent
concern. Hook had some critical things to say about the proposed new
Western Culture course at Stanford—one that is designed (in the words
of the syllabus) "to emphasize the contributions of women and minorities
to Western culture," (not that these have been neglected, you can be
sure!), and includes the study of such figures as Frantz Fanon and Herbert Marcuse.

Sidney said that he was surprised to find professors in the humanities
"yielding to the political demands of those who are not familiar with the
classical liberal traditions of the West." Professors are supposed to be "the
guardians of the Western tradition," he said, but today increasingly "they
capitulate—they are afraid of the women and the blacks, especially the
blacks. I never thought I'd live to see it!"

He said that this tendency on the part of the academy was a recent thing. "There was a time when the professoriate was, if anything, hostile to the mildest kind of social reform," he said, thinking of the time in the 1920s when he first joined the academic world. The change in values didn't begin until Franklin D. Roosevelt's time, he said.

I asked him to what he would attribute the weakness of the West in recent decades. "Some would say the erosion of religion," he replied. "But that doesn't explain why religious belief eroded." Sidney commented that many traditional conservatives believe the French Revolution was "the second Fall of Man." Not so, Sidney believes. It was World War I, an unnecessary war and one that could have been settled amicably; instead it produced such a terrible holocaust. Ever since then, he feels, fear of war has undermined traditional values in the West. "That fear has really eaten into the vitals."

He mentioned the view once held by his friend Bertrand Russell, that "we could have used the atom bomb skillfully against the Soviet Union," at a time when they still didn't have it. "But we didn't," he said simply. When he said this he seemed almost wistful, as though contemplating an opportunity missed, but later he made it clear that he considered Russell's position "rash and unnecessary," making him "morally unfit" to sit in judgment on President Kennedy during the Cuban missile crisis.

I asked Hook if he still called himself a socialist, a philosophy in which he had obviously invested much intellectual capital earlier in life. "If you let me define it as a belief in democracy as a way of life," he said. "Socialism without freedom is slavery."

A puzzle with Sidney Hook is to understand why he has consistently been so uninterested in Israel. To a Jew, was this not the great miracle of the 20th century? "As young socialists we had illusions," he told me. "We didn't want to be nationalistic. So I was never a Zionist. Then, after the state of Israel was created by the United Nations I had to accept it." It was of course the universalism of socialism that had appealed to him. But for him the utopian vision of socialism had obviously faded, and evidently he had never felt the need to substitute another faith for it.

By now we were approaching his home near Mayfield Avenue. He said that he would have to have an operation soon because the doctors had told him he had "the same thing the President has." He said he wasn't afraid of the operation.

Without my asking, he told me that a friend once asked him what he would say if he woke up in the presence of God. "Lord," he would say, "you didn't give me enough evidence."

In the Castro

The San Francisco Chronicle reported that "prostitutes and drag queens cavorted in skivvies on one side of the street in front of Mayor Dianne Feinstein's house last night. Nazi death camp survivors and Orthodox Jews chanted holy songs on the other."

They were protesting the mayor's $250-a-plate fund-raiser for Pope John Paul II's two-day visit to the city in mid-September, 1987. Another paper reported that the 50-odd demonstrators were "more anti-Pope than anti-Feinstein." The Jews were there to express "outrage" over the Pope's audience with Austrian President Kurt Waldheim. The "drag queens," a group of about 20 homosexuals who dress up as nuns and call themselves the Sisters of Perpetual Indulgence, are outraged because the Pope has called homosexuality an "intrinsic moral evil." As for the prostitutes, Scarlet Harlot, leader of the Whores of Babylon, sang a tune entitled "Pope, Don't Preach, I'm Terminating My Pregnancy."

"Use Condoms," say the ads on San Francisco's buses. The AIDS hotline number is appended. About 3500 cases of AIDS have been diagnosed in San Francisco, with over 2000 deaths to date. About 97 percent of the cases involve homosexual or bisexual men. In June, 96 new cases of AIDS were diagnosed in the city, 95 of them homosexual men, six of whom also used intravenous drugs.

A newsclipping attached to the notice board of the First Unitarian Church quotes F. Jay Deacon of the denomination's Office of Lesbian and Gay Concerns as saying that the Vatican's linkage of AIDS and homosexuals is "mischievous and arrogant." The Unitarians believe the Vatican's 1986 letter on homosexuality is "laced with archaic religious assumptions and astonishing arrogance."

Also on the Unitarians' notice board: an advertisement for the Hemlock Society, and an announcement of a forthcoming meeting of the "Womyn's Spirituality Circle." (Witches, they used to be called.)

Estimates vary, but there are perhaps 75,000 homosexual men in San Francisco—one tenth of the city's population. According to one widely quoted figure, half of them may be infected with the human immunodeficiency virus that causes AIDS. Many of these could contract AIDS.

A local ordinance prevents San Franciscans with AIDS from being fired from their jobs. Nor can they be denied or evicted from housing simply because of an AIDS diagnosis. Insurance companies in California are not allowed access to results of the AIDS antibody test in assessing insurance premiums. There was even a case in Los Angeles where a judge ruled that AIDS was a job-related illness and awarded worker's compensation payments to a man who said he had contracted AIDS from prostitutes in Zaire.

It was only in May that the last homosexual "bathhouse" in San Francisco was closed, and that after much grumbling and accusations of official harassment. In the late 1970s there were 30 such establishments—dimly lit rooms designed for anonymous homosexual encounters. The lawyer for the last one to close said that the two owners plan to retire, having in their own estimation "provided a service to the community." But they added that "the future for gay bathhouses at the moment is not very bright."

Everyone says that the Castro district of San Francisco, predominantly homosexual (and to an increasing extent also lesbian), is much more subdued that it was at its heyday in the late '70s. Today you see slim, prematurely aged, greyish, unshaven men leaning on canes as they come creeping down the streets. But the bars on Castro Street still have a good many customers, even in mid-afternoon, and on weekends they're packed. *The Bay Area Reporter*, catering to homosexuals, still comes out with pages of more or less unprintable classified ads, and pictures of nearly nude men, frequently on the telephone. ("Tired of the same unbelievable *fantasies?* You've never had a sex call this *hot, nasty* and *sexy*. . . .")

Turn the page and you come to *Deaths*—perhaps a dozen obituaries every week, almost all resulting from AIDS. "Due to an unfortunately large number of obituaries," the paper explains, "*Bay Area Reporter* has been forced to change its obituary policy. We must now restrict obits to 200 words. And please, no poetry." Many of them are written by a "lover," who may of course have transmitted the virus to the deceased.

About 200,000 San Franciscans turned out for the annual Gay Pride Parade down Market Street in late June. "Whips were everywhere," according to the *Bay Area Reporter.* "A stand-in for Pope John Paul II rode in a pope-mobile pulled by the Sisters of Perpetual Indulgence. 'His

Holiness' was garbed in splendid pontifical robes and carried a whip with which he threatened the sisters whenever they looked like they were beginning to get lax in their duties."

Local politicians cannot afford, apparently, to miss this event. Everyone from the sheriff to the district attorney turned out for the parade. In the current mayoral race (Feinstein is not seeking reelection) a surprising issue emerged between the two leading contenders when state Assemblyman Art Agnos carelessly said he "didn't have time" to campaign in "leather bars," frequented by those addicted to sadism and masochism. This remark was interpreted as an attack on his opponent, San Francisco Supervisor John Molinari, who had found time for such appearances. Agnos then found himself on the defensive, denying that he was in any way homophobic.

Only one candidate has dared raise questions about homosexuality, and he is a political newcomer who doesn't know the ropes. "He said 'Some of my best friends are gay but I wouldn't want my daughter to marry one'— that kind of nonsense," said Gilbert Block, who plays Sister Sadie the Rabbi Lady in the Sisters of Perpetual Indulgence. "If he gets more than one percent of the vote we'll all be *very* surprised."

The sisters are planning various protest activities and street theater for the Pope's visit. Some of them no doubt will be wearing Papal robes and there have been pictures in homosexual publications of terriers in mitres. Sister Boom Boom (real name Jack Fertig), whose demonstrations against Phyllis Schlafly and Jerry Falwell of the Moral Majority gained him fame or notoriety at the time of the 1984 Democratic National Convention, will no longer be with the group. He told me that in 1985 he gave up drugs and alcohol, and soon thereafter began going to church "seriously." Remarkably enough, at the time of the 1984 demonstrations Archbishop John Quinn of San Francisco chose to criticize Falwell and Schlafly, rather than the homosexuals who were using nuns' habits to ridicule them—a perhaps extreme example of the American Catholic hierarchy's recent eagerness to curry secular favour by going along with any and every trend.

I attended an anti-Papal fund-raising event given by John Wahl, a "gay rights" lawyer whose Papal Visit Task has been mobilizing a coalition "to express the united outrage of Californians at the Pope's public condemnation of gay liberation, women's right to choice, and liberation theology in Third World countries." It was held one Sunday afternoon at the Langtry Inn, a Victorian guest house for women travellers partly owned by Ms. Ginny Foat, the former California president of the National Organization of Women, who was acquitted of murder charges in Louisiana in 1983. Foat said she was a practicing Catholic but regretted that the Pope was "not a compassionate man," ideally in her mind a "combination of Martin Luther King and Gandhi."

Also on hand was Robert Tielman, something of a celebrity and introduced to all and sundry as the organizer of the anti-Papal protests in Holland last year. He told me he was "head of the Gay and Lesbian Studies Department at the University of Utrecht," and he wanted me to know how deplorable it was that this Pope was "imposing his conservative views on liberals in the church."

"That man is coming here in September," said John Wahl, who was attorney to Harvey Milk, the homosexual San Francisco politician shot in City Hall in 1978 by Dan White (a "closeted gay man," according to Wahl). "John Paul wants to walk our streets like a gentle grandfather, patting the heads of supplicants. Is that what you're going to do?"

The audience in the sunny, chintz and floral patterned Lillie Langtry drawing room murmured no, no, no.

"That's not what I'm going to do," said Wahl. "That's not what thousands of people are going to do. We're going to tell that man and every other so-called religious leader that says that kind of thing, that is guilty of that kind of lie, that kind of inflammatory rhetoric, we're going to say to him or her . . . go home, go back to bed, go back and get psychoanalysed, get some therapy, you're sick, you're impeding the progress of humankind in advancing and growing up."

While I was in the city there was an extraordinary tribute to Dr. Tom Waddell, the founder of the Gay Olympics, now called the Gay Games in the wake of a Supreme Court copyright ruling. Dr. Waddell recently died of AIDS at the age of 49, leaving a wife and three-year-old daughter. He was eulogized in the rotunda of City Hall in terms that consciously evoked the praise heaped on Robert F. Kennedy after his death. About 400 people gathered to honor Waddell for his "heroic life and the legacy of inspiration it has left them," according to *The San Francisco Examiner.*

The tribute was actually organized by the city's Department of Public Health, for which Waddell worked as chief physician until he came down with AIDS himself. I couldn't help thinking as I stood in the solemn crowd: had the public health doctor transmitted AIDS to others?

There were mauve cummerbunds, women in men's clothing (blazers and ties), U.S. Naval officers in whites and medals, a good many young men with floppy mustaches, and a solemn row of eulogizers awaiting their turn at the podium. "Gay rainbow" flags were on display. A pianist played "Look for the Silver Lining," and there was the usual sign-language accompaniment (increasingly provided with or without deaf people in the audience).

Health Commissioner James Foster told us that Waddell wanted to heal not only the diseases of the body but also the soul—diseases that "divide gay from straight, men from women." He wanted to break down "walls of oppression." And so on. "We in Public Health are grateful . . ." I heard him say. We heard from the director of the Health Department, who told us that Waddell was a caring person. We heard from the mannish Mary Dunlap, lead counsel to San Francisco Arts and Athletics, who deplored "this disease which picks on our leadership and our fellowship in the most insidious way." And we heard from the widow, Sara Lewinstein, a Lesbian in tailored jeans and dressy high-top Adidas who had met Waddell at the first Gay Games. "The actual vows of the marriage ceremony didn't mean anything to us," she said. They had lived apart. "There are men and women who want to have a child, and their sexuality has nothing to do with it," she said.

Finally we heard from the associate Director of Health, Dr. Tom Peters, who said: "You have provided us with a model for years to come. As we train new health care providers, we at the Health Department will always remember you, Tom."

"Ladies and gentlemen, the Lesbian and Gay Chorus," the announcer said, and they came solemnly down the City Hall staircase while the piano played "Hand in Hand."

"Women in drag sitting in the rotunda of City Hall memorializing a man who died of a venereal disease," said my friend Charlotte Hays as we left, "and no one seemed to think it was odd."

Back in the Castro district, I spoke to a Catholic priest who was coming out of the Most Holy Redeemer Church and who visits dying AIDS patients in the "Coming Home" hospice opposite the church. About 40 men have died there since it opened earlier this year. I asked the priest, Fr. John McGrann, if the dying men asked for the Sacraments. He seemed shocked that I would ask such a question. One had, he admitted. Fr. John told me that he had earlier left the priesthood and had run a flower shop in the Castro district for four years. But Archbishop Quinn had readmitted him to the priesthood.

"The Pope is in for a big surprise when he gets here," Fr. John said. "People are hurting and he comes out with statements that are harsh and judgmental. The church is not a very loving place." He said the Pope should "stay home and send us a videotape."

In the Land of Lenin

There were twenty of us in the Finnair Lounge at John F. Kennedy Airport, preparing to depart for our ten-day tour of the Soviet Union. We would be visiting Leningrad, Tbilisi and Moscow, in that order. Tbilisi used to be called Tiflis, and it is the capital of Georgia. Ernest Lefever of the Ethics and Public Policy Center asked me if it was my first trip. I said yes and asked the same question of Douglas MacArthur, a former U.S. Ambassador to Iran and Japan, now retired from the Foreign Service.

"No," he said, "I went to Moscow in August 1939 with a letter from Roosevelt to Stalin. We had heard that the Ribbentrop-Molotov Pact was about to be signed. Roosevelt wanted to dissuade Stalin because he knew it meant war. But we were too late. When I arrived the pact had just been signed."

I thought this rather a good answer to a getting-to-know-you question. It almost might have been a put-on, so historic were the events he casually associated himself with. But he was perfectly serious. Douglas MacArthur had the doleful eyes of a basset hound, and the high balding dome of his famous uncle.

"Did you meet Stalin?" I asked.

"Not on that occasion."

Doug, as we took to calling him, acted as our unofficial chief of protocol, toastmaster and speechmaker. The leader of the group was Larry Moffitt, a 36-year-old Texan with a ready repertoire of laconic phrases. A member of the Rev. Sun Myung Moon's Unification Church, he was a veteran of two earlier Soviet tours. Among those also making the trip with us were Arnold Beichman, a former newspaperman and professor, now a columnist affiliated with the Hoover Institution; Richard Brookhiser of *National Review*; *Washington Times* foreign editor Holger Jensen; and Dixy Lee Ray, the former governor of Washington state. The tour was arranged and paid for by News World Communications, the organization that founded *The Washington Times*.

As we circled above Leningrad I looked below for signs of life but could see little more than desolate, snowy terrain: a lonely road with a lonely truck, a lonely winding river. This was early April but from the air it might have been February. There was no suburban development as we understand that phrase. Leningrad Airport's cracked and rutted runway would scarcely have passed inspection by the Federal Aviation Administration. The air quality (aviation fumes, grit, strong cigarettes) would not have satisfied our Environmental Protection Administration.

Customs was an ordeal. Officials made off with a copy of David Shipler's innocuous book, *Russia: Broken Idols, Solemn Dreams*, brought in by Martin Nolan, the editorial page editor of *The Boston Globe*. They promised to return it to him when we left Moscow, but did not. They took from my briefcase a research report on Yurchenko, the re-defector, and kept it; but let through (with a second glance) a book of homilies by Msgr. Josemaria Escriva, the founder of Opus Dei.

The customs inspection of our group was wholly ideological. They took no interest in our material possessions, a lot of interest in the printed matter we brought in with us. Suitcases were ignored; briefcases searched. Uniformed guards leafed slowly through two news magazines I had brought along, pausing to scrutinize photographs of the USS Saratoga in the Gulf of Sidra. An interview with Jeane Kirkpatrick, reproduced from *Encounter*, was solemnly transported away to some higher authority, with my passport as hostage, and grudgingly returned half an hour later.

Olga, our Intourist guide, a member of the Communist Party, watched noncommittally from the sidelines. Larry Moffitt said that he had never seen an Intourist guide tell an immigration official what to do. (Establishing these hierarchies of command, determining who is subordinate to whom, most be one of the most difficult tasks in setting up a Communist state.) Olga seemed to belong in a spy movie. She wore a maroon pork pie hat, high boots to match, and belted raincoat with collar turned up. She spoke good English and kept a comradely eye on us throughout our tour. Poor Olga was forever filling out forms in duplicate and triplicate, fussing with little bits of carbon paper like an old-fashioned shopkeeper, and I believe she had to spend many a night filling out more forms that we never saw—about us. (This from Intourist guides who have defected.)

We were, incidentally, free to leave the tour at any time, and many of us did so. But what do you do and where do you go in a strange, foreign land where you don't even understand the alphabet, let alone the language? Mostly we just stayed with the tour.

Our first glimpse of socialist construction came as we drove by bus from airport to hotel down a lightly traveled road—this the *main* road from the airport to the city center. On both sides of us lay muddy, empty wasteland. (I say wasteland rather than empty lots because "lots" implies property and boundary lines, both long ago outlawed in accordance with the first principle of Communism: the abolition of private property.) Haphazard mounds of earth dotted the roadside, along with splintered timbers, rusted pipe, occasional evidence of half-hearted spadework, long since abandoned. ("The state protects socialist property and provides conditions for its growth," says Article 10 of the Soviet Constitution, English-language copies of which were available in our hotel. "No one has the right to use socialist property for personal gain or other selfish ends.")

Our hotel, Swedish-built for the 1980 Olympics, was on the outskirts of Leningrad. In the normal course of events Soviet citizens were not allowed inside. The Communist authorities labor under a great and permanent dread that Russians will meet foreigners and thereby learn forbidden things. This is remarkable since most Soviet science, technology, architecture and (especially) ideas seem to have been imported, and since the 18th century this has been true. Sometimes in the course of our tour it seemed that the Russian language and the Russian Orthodox Church alone were indigenous.

In the vast hotel lobby was to be heard the background sound of a fast-paced beat—Leningrad rock, as I came to think of it. More Western influence, and how I wished they *had* kept that out. Later, in Moscow, we told Olga that U.S. tourists would flock to a nightclub where Tolstoyan peasants strummed balalaikas—a Russian version of New Orleans's Preservation Hall. Olga was unsure whether to take this as a compliment to Russia or an insult to the very Western "floor show" we had just watched. I think we had simply wearied of imitation. I suspect also that Olga may have thought us unprogressive and sentimental at heart—no doubt true enough.

Our hotel rooms were okay—light bulbs almost too dim to read by, however. Larry had warned us not to drink the water in Leningrad. The bottled mineral water was safe, he said, its drawback being that "it tastes as though it has been run through a carp." One mouthful provided adequate confirmation, and thereafter one was reduced to Pepski.

*   *   *

Six of us stood outside the hotel in the long twilight. Opposite stood a memorial to the Great Patriotic War (World War II). Across the way blocks of dim, depressing flats disappeared into the distance, bringing to mind the ambience of George Orwell's *1984*.

Arnold Beichman, a short, stout man with a white thatch of hair, was talking about one of his favorite subjects, Vladimir Ilyich Lenin, whose icon was everywhere. (I hardly remember seeing one picture of Mikhail Gorbachev.) Beichman was saying that Lenin was the only "historical figure" in the Soviet Union, the rest being confined to the shadows. "Where are the collected works of Brezhnev today?" he asked. "There are historians here who devote their time to studying a week in the life of Lenin."

Beichman said he had never before been allowed into the Soviet Union. It seemed from his conversation that he had spent half his life reading, thinking and arguing about the country. His father had emigrated to the U.S. from the Ukraine, he told me. Now in his early seventies, Arnold had grown up in an orthodox Jewish household in New York City. He was supposed to become a rabbi, but somewhere along the line he read H.L. Mencken, he said, and today he is an agnostic. Even as a young man at Columbia University in the 1930s he had no illusions about Marxism (thanks to his father's influence), and then, as now, he spent much of his time battling the many who did.

He mentioned Lenin's 1902 pamphlet, "What Is to Be Done?" This he regarded as one of the most influential documents of our time—a prophetic "blueprint" of the 20th century. Here Lenin warned that the proletariat would never bring about revolution on its own because it would always be bought off with crumbs from the capitalist table. It would need help from the true revolutionaries—disaffected intellectuals of the propertied class.

On the spur of the moment we decided to go to the center of Leningrad on the Metro (opened in 1955). I was not prepared for the dense crowds swarming on the sidewalk, but perhaps I should have been—the population of the city is almost five million. The riders coming up the escalator of the immensely deep subway looked curiously across at us. Foreigners! Wladyslaw Pleszczynski, the managing editor of *The American Spectator*, said how much dirtier the subway looked than when he had lived in Leningrad for a few months in 1977.

"I just realized," Beichman said on the train, "all these people are Communists." We looked about us with renewed curiosity. "I've spent all this time criticizing the Soviet Union and now I'm beginning to worry that nothing iniquitous is happening," he added. The people were law-abiding; no guns to their heads; going home to wives and children; some with packages or books in their laps. No muggings underway either. Beichman said that his immediate reaction was obviously also the reaction of many travelers to the Soviet Union. They see little to fear, regard anti-Communism as paranoid, and so perhaps set off on peace marches: Russian humanity disguises the political inhumanity. They're just like us! (So why can't we live together in harmony?)

Still, it did make an impression to see the Russians who daily had no choice but to submit to the Communist system of command from above. "I think Gorbachev knew what he was doing when he let us in," Beichman said.

Nevsky Prospekt was packed with people trudging along in the dim twilight. Not much streetlighting. Not many cars. Not much in the shops, if you could even call them shops. Everything looked run down and dilapidated. Nothing had been swept or painted in years. Occasionally a young man would dart up to us and offer to exchange roubles for dollars at an exchange rate closer to the illegal market rate.*

Wlady said that in his 1977 stay in the Soviet Union he was never so approached. Perhaps there is change here, perhaps a growing fearlessness on the part of the young, who seemed on the (untrustworthy) surface to be more "capitalistic" and more independent-minded than their elders. On the other hand, it is possible that this trend has been observable in the Soviet Union for decades and has always led nowhere.

A youth did a double-take as we breezed along Yankee-Doodle style. Out of the corner of my eye I saw him break away from his companions and approach us: another hopeful money-changer. (We refused all such offers because, of course, they are illegal and could have landed us all in trouble.) The young man spoke good English. He walked along with us

---

*The official exchange rate is 7.2 roubles for $10. Freelance money changers regularly offer 3 roubles per dollar, and one heard that the real "market" rate is about 5 roubles per dollar. Thus Soviet officialdom intercepts perhaps 85 percent of all U.S. dollars spent in the Soviet Union.

Different sources give different figures, but perhaps 60,000 U.S. tourists visited the Soviet Union in 1985. The Helsinki Commission thinks that this figure is too high, claiming that only 35,000 Americans visited the USSR in 1984. If we assume that these tourists spend on average $1000 per person, U.S. tourism yielded the Soviet government approximately $50 million in clear profit last year. This, however, is a small fraction of the hard currency that the Soviets now stand to lose as a result of the oil price drop.

for a few blocks. He was openly scornful of Communism and implied that none of his friends took it seriously or believed a word of it. He had read Hedrick Smith's book *The Russians* and thought it accurate, but in general he found it hard to lay his hands on books in English, other than innocuous Eng. lit. books. Afghanistan? He was worried about that because he knew he would be drafted and quite likely sent there. Recruits from regions bordering Afghanistan (whom he referred to as "Arabs," meaning Muslims) had proved to be unreliable, the Soviet authorities had found.

Beichman asked the young man if he was worried about being followed. No, he said, the authorities would only be keeping an eye on us if we were White House big shots or, let's say, notorious anti-Communists. I didn't have the heart to tell him that Beichman was one of the most notorious in the United States. Arnold slipped him a few dollars and he was gone in a second.

The Winter Palace loomed up in front of us. One recognized the scene from those mesmerizing photographs taken 70 years earlier. All was quiet within the well preserved palace, the large square in front of it deserted.

"Kerensky, that jerk," Beichman said. "One bullet could have changed history. A lot more people would be alive today."

I demurred, taking my cue from *War and Peace* (in which Tolstoy argues that Napoleon himself is insufficient to explain events of the magnitude of the Napoleonic Wars). The Russian Revolution was too big an event to attribute to the genius, evil or otherwise, of one man. It would all have happened anyway, even if Lenin had been strangled in his crib, let alone shot by Kerensky. We argued about that, just as we argued later about the Pipes v. Solzhenitsyn dispute (the former blaming Russians, the latter Communism).

We walked the short distance to the Neva River. White ice floes slipped through black water. Across the river was the Peter and Paul Fortress where Dostoevsky was imprisoned and, someone said, Bakunin convinced his jailers to become anarchists.

"I'm sorry, but it does have an effect to be here," Beichman said. "Where's the monster? Where's the beast in the belly?"

"Arnold, you've given in already," said the columnist Richard Grenier.

"There's the Seine, there's the barges," said Richard Brookhiser, teasing Arnold.

The next day we went sightseeing on the Intourist bus. Richard and Cynthia Grenier, sitting as was their custom in the front seat, kept asking

Olga to translate the hortatory slogans on top of many buildings. ("The Goals Of the 27th Party Congress Must Be Fulfilled!") Olga clearly did not enjoy having to furnish us with these Communist commercial breaks. Indeed, overall our tour was almost completely non-ideological. No factories, no collective farms, no Museums of Economic Achievement. The Intourist goal seemed to be to minimize the idea of a revolutionary state.

In a way Leningrad is a preservationist's dream. There are many fine eighteeenth- and nineteenth-century buildings in the center of the city (the twentieth-century outskirts are unspeakable, however), but they badly need repairing. Some have been kept up, but in general the problem of maintenance far outruns the capacity of a Communist society to cope with it. While one building is being restored, ten more deteriorate. Everything looks shabby in socialist countries because in the absence of private property only top priority things get done—servicing the military, putting some food on the table. Sprucing up buildings can never be high on the central planner's list of priorities.

Leningrad—St. Petersburg, as it used to be—escaped the destructive advance of both Napoleon and Hitler (although the German army came within a few miles of the city). Perhaps as important, many old structures have survived because the Communist state today scarcely has the surplus energy needed to tear down old buildings, let alone put up new ones in their place. For that reason, Communist countries look like dilapidated museums of decades past—something that has also been observed of Havana. This is so comically at odds with the old idea that Communism represents the future, the "new society," that I found myself wondering anew at the strange persistence of socialist ideology among intellectuals all over the West.

. . . Without their succor, their aid and comfort, above all their faith, this society wouldn't last five minutes . . . I found myself thinking more than once while I was in the Soviet Union. And that perhaps is the real crisis that Gorbachev must now face—not so much restless Soviet consumers as restless Western intellectuals, more and more of whom have been over into the Soviet Union and found not a working future but a ramshackle past.

We stepped out of the bus for a lunch that was not so great, but the bread was fine, and plentiful.

After lunch we were taken to St. Isaac's Cathedral—man cannot live by bread alone. Beichman said: "Tourism, like diplomacy, is easy on the head but hard on the feet." I believe this was half cribbed from someone else but still it wasn't bad. St. Isaac's, a Russian Orthodox church not now functioning as such, had been elaborately restored at considerable expense to the state. The mosaics of Christ and the apostles had clearly been worked on with loving care. And it was the same in so many churches that we visited. We saw so many churches, monasteries, and icons that one might at times have thought it a tour of Holy Mother Russia. And as someone remarked, much of what our Intourist guides told us would have been considered "unconstitutional" in the U.S., violating the separation of church and state.

I asked Olga why a Communist state would spend so much money restoring the symbols of Christianity. "Because it is part of our past," she said, "and without the past there will be no future."

This was a pat answer, no doubt oft-repeated to tourists. In part it acknowledges Communism's failure to produce anything new. Without the things of the past that Communism has chosen not to destroy there would indeed be nothing. Still, the decision of the Party not to destroy all public traces of Christianity is surprising, considering that Communism is itself an inverted religion, rooted in antagonism to all other religions (especially Christianity). There are about fifteen functioning churches in Leningrad, and (we were told) forty in Moscow. (Not many, per capita, of course.) These churches and their interiors, to say nothing of what takes place inside them, are so much more impressive and beautiful than anything that Communism has produced that one would have thought the contrast subversive; far more unsettling to Communism than anything U.S. tourists might bring in their luggage. But one thing I did notice about the (functioning) churches I visited: There was no written literature available in any of them. No hymn books, prayer books, church bulletins. Everything must be recited from memory. Evidently the Communists believe that an aurally transmitted doctrine poses minimal risk. Their faith in the written word is boundless (betraying the intellectual rather than proletarian roots of Communism); hence also their devotion to literacy campaigns, as Lenin himself admitted. (The proletariat must learn to read, he said, so that they could understand, and obey, Party instructions and slogans.)

That evening we heard a Donizetti opera at the Kirov Theatre. Keeping us up to the historical mark as usual, Beichman reported to the group that S. M. Kirov had been assassinated by Stalin in 1934, thereby ushering in the Great Terror.

"Stalin always did in his Leningrad bosses," Beichman added as the curtain went up. It was something to keep in mind, certainly. I told him that we were about to watch a commie opera. He said that would be enough Red-baiting for the evening. As usual at the opera I found that I couldn't follow the action at all and soon dozed off, my head rolling over to one side, thinking dim thoughts about the ornate dress circle, ornate building, ornate opera, all the color and pageantry of the old order intact and permissible under rubric of culture, all very non-Communistic. . . .

Someone suggested making a list of the outward and visible signs of a Communist society—queues, slogans, portraits of the leader, and so on. The absence of pornography was one suggestion. This of course should characterize a Christian society. Another sign is the Beriozka: the special store where only hard currency is accepted and from which workaday Russians are excluded. There are many such luxury stores in the Soviet Union today, some for tourists, some for Party officials, and so on. They mock the Communist ideal: luxuries for the privileged. They have existed from the beginning of the Soviet state—the Communist equivalent of the sale of indulgences.

Can reformation be far off? The Beriozka is subversive. The ordinary Russian must have noticed by now that these special shops are patronized by two groups of customers: Party officials and foreigners. The reasonable inference is that *both* are privileged by comparison with the Communized proletariat. The Beriozka tells the perceptive Russian that the Communist masses in all probability have nothing to lose but their chains.

Nonetheless, we were exhorted to visit the Beriozka at every opportunity, so that we would leave our dollars behind. The Soviet Union is probably even more "infiltrated" by the market than is the U.S. by power-hungry intellectuals dreaming of a new socialist society. No doubt Gorbachev would like to restore his domain to its pristine Communist state (assuming it ever existed), but there is no way that he can do so. The Party elite will insist on retaining its legal privileges; and beneath them there is the substratum of "corruption": the illegal underground economy, recently estimated by Vladimir Bukovsky at 30 percent of the total production of goods and services. Party authorities are under relentless "capitalist" pressure to turn a blind eye to such illegal exchanges. If they "crack down," then material hardship will ensue. But if they ignore this illegal trading, then prosperity will increase and the people's dependence on Party control will diminish.

It was the consensus of those in our group who had been to the Soviet Union before that the material condition of the Soviet proletariat has noticeably improved in the past decade or so. (And in no country that I have visited is there more conspicuously a proletariat.) There is less waiting in line for essentials, and items such as fresh vegetables are more readily available. Clothes are brighter and better made.

As a newcomer, nonetheless, my first impression was of the tremendous disparity between the general level of economic development in the Soviet Union and the U.S. The real Soviet GNP (sometimes said to be about 60 percent of the U.S. level) is undoubtedly a small fraction of ours—perhaps no more than a tenth. That is a pure guess, but there can be no doubt that the official Soviet GNP is greatly overestimated.

I kept thinking how extraordinary it is that for 70 years the regime has managed to preserve, semi-intact, a materialistic ideology that has so conspicuously failed at the material level. Still, the point is that this is only conspicuous to those with direct experience of a capitalist country. The material condition of the Russian people has slowly improved, after all. As long as they are kept in the dark about conditions in other parts of the world they might reasonably suppose their condition to be normal. All the more reason, then, for their rulers both to undermine capitalism wherever possible and to limit the proletariat's contact with capitalists.

Before we left Leningrad, Arnold Beichman visited the only synagogue and returned to say that the regime had made important concessions in the realm of religious practice. "Keep your nose clean," Beichman said by way of summary, "don't apply to emigrate, don't agitate for Israel, and they'll let you alone." The regime's attitude seemed to be: "Let them eat matzos." Unleavened bread was available for Passover. Kosher arrangements could be made. The observant Jew's requirements had been catered to.

"So why make trouble?" Beichman asked. "Does everyone have to be Scharansky? It raises the question of what you would do in a totalitarian society. Let's say you have got a wife and a couple of kids. Are you going to be a character witness for someone who has been arrested? What's in it for you?" To fight Communism in New York City during World War II took heroism, he thought. "Still, you weren't going to be sent away for 20 years." In short, he didn't believe that the Jewish rebellion in the Soviet Union would grow, given the regime's present tolerance of Jewish practice.

I had visited the (Polish) Roman Catholic Church that morning and felt much the same way. Don't proselytize, don't make a fuss, and you can probably go to Mass without getting into trouble. (Of course, you could not do this if you were a member of the Communist Party or wanted to join it.)

I made my own way to the Hermitage Museum, which is inside the Winter Palace. But I never could find Olga and the gang in that vast, baroque maze, so I strolled about on my own. At one point I sat in a large, conservatory-style room, surrounded by old paintings and well behaved Russians and experienced the sense of peace and contentment that other fundamentally unsympathetic visitors to the Soviet Union have also experienced: a feeling of release, however temporary, from media strife, ideological conflict, porn and pot, accusatory minorities, feminist nuns (bewitched and bewildered), batty bishops, haywire academics, and the entire panoply of Western lunacy, no part of which would be permitted to survive for five minutes in the Soviet maw. . . . Oh, I know, it's all a part of the price we pay for living in a free society. . . . But is it *really* necessary? Dare one point out it's all very recent. . . .

I moved on and found myself completely alone in a smallish eighteenth century room, and to a degree that is unusual in museums was encapsulated in the past: a museum in a historic building in the midst of a city itself mummified by Communism. There was no museum guard in sight. My feet aching, I sat down in a gilt and velvet chair next to a marble table. Voltaire and his friends might have come in at any moment.

Footsteps! The heavy tread of guards. I looked up with alarm, expecting to be thrown out for breaking the rules. Two soldiers of the Red Army came into the room, squat and bulky both. They took no notice of me as they walked through the gallery . . . arm in arm. Like a courting couple! They were obviously confident that their apparent intimacy wouldn't be misconstrued. Unthinkable in this country! I wondered if they would soon be off to Afghanistan, and recalled the cloud of anxiety that crossed the face of the youth on Nevsky Prospekt.

Perhaps the children of Czar Nicholas II had played in this very room. I looked out of the window to see what they would have seen, and there was the Alexander Column, put up in 1832, topped still by the angel holding the Cross of Christ.

In Leningrad Airport there were soldiers and airmen everywhere. The country seemed to be in a state of continuous mobilization. We were kept waiting for our flight to Tbilisi. In the Soviet Union you soon learn to wait patiently. Nearby was a group of Syrians with a caged bird.

Our Aeroflot jet seemed to have seen recent military use. We were packed in uncomfortably tight—knee to kidney. In pitiful mimicry of

Western ways, an attendant once during our three-hour flight brought around a tray with clear plastic cups of metallic-tasting water.

I sat next to Larry Moffitt, who told me a little about the Unification Church. Members apparently believe that Communism fulfills the Biblical prediction of the Antichrist. This was not so very different from Whittaker Chambers's apocalyptic foreword to *Witness* (1952), in which he wrote: "I see in Communism the focus of the concentrated evil of our time. . . . Within the next decades is to be decided for generations whether all mankind is to become Communist." And Comrade Gorbachev said at the Party Congress in February: "This is perhaps the most alarming period in history."

I could see little other than haze from the window. How our satellites know where to look for their mobile missiles is a mystery to me and I suspect to the Pentagon. Presently the icy terrain of the Caucusus came into view and we seemed to fly quite close to one tremendous peak. One had heard horror stories about Aeroflot. Lev Navrozov said that when a plane crashes they simply bulldoze earth over the wreckage and nothing is reported in the newspapers. So it was a relief to touch down in Tbilisi, brightly lit in a valley between the mountain passes.

Georgia, USSR, is has about the same population as Georgia, USA. One was tempted to think of Tbilisi (pop. 1.2 million) as their Atlanta. Georgia has its own language and alphabet, which the Soviets tried to suppress a few years back. This could be seen as the reverse of our own dotty anti-melting pot efforts—funding bilingual education, preserving Cajun French, and so on. But the Russians found themselves with a Georgian rebellion on their hands and so they abandoned their exercise in cultural imperialism.

The main street, named after a Georgian poet called Rustaveli was brightly lit. (Why are poets so worshiped in our socialist age?) There were sidewalk flowerbeds and roadside trees. The city had a dusty, Moorish, Mediterranean air—most agreeable after Leningrad's muddy battleship grey. We all felt much more cheerful. The next day Rick Brookhiser said, I believe accurately: "We have come to a place that is culturally and economically on the level of Turkey, and we are overjoyed."

Our hotel, the Iveria, was built in the early '70s—a somewhat ramshackle but quite pleasant 20-story tower with Hilton-ish aspirations. The doorman gave us a non-Communist bow as we entered. Civility was still alive south of the Caucusus, apparently. Interior hotel signs were in English only: Service Bureau. Night Bar. Currency Exchange. At the Service

Bureau the next day a radio was playing "Chattanooga Choo Choo." Strolling yanks wore pastel pants.

Imagine you went to the main hotel in Atlanta and all the signs were in Russian. And only in Russian. What would you think? And imagine that illegal entrepreneurs on the streets were trying to spot Russian tourists so that they would change dollars for more desirable roubles. Teenagers sought out Russian clothes and danced to Russian music. On the radio: Hits from Moscow.

I could see why the Communists think capitalism is imperialistic. It really is—at least at the level of commerce. Of course it leaves intact the machinery of government, but it does bring pressure to bear at the more fundamental level of everyday life. At the 1986 Party Congress in February Gorbachev noted with alarm that capitalism's "productive forces" have "grown to gigantic proportions." The result is that non-market countries trying to preserve their regimes of privilege against the forces of property are under increasing strain. In this sense capitalism is inherently anti-Communist. International capitalism subjects Communism to intolerable competitive pressure—or would if it did not simultaneously pay tribute to power in the form of subsidies.

Describing her recent stay in the Soviet Union (Tbilisi in fact) Stalin's daughter Svetlana told the *New York Times*: "You could meet a taxi-driver, or a man selling vegetables at the bazaar, suddenly bursting forth into talk about how much 'private property is needed,' or 'nothing can be done without private initiative.' I was surprised how people began to talk about that, without even being asked. It must have been on everyone's minds." It was certainly on mine.

Larry Moffitt said the more you move away from the center, that is from Moscow, the more attenuated Communist power becomes. Tbilisi seemed to be a good deal freer, more productive, more relaxed, not to mention more congenial than Leningrad. There were more cars on the streets and more goods in the shops. Obviously there was a fair amount of *de facto* capitalism, or free exchange. And a Party crackdown would only undermine the productivity upon which Moscow depends. Wherein lies the dilemma of the Communists: Let them produce, or keep them under control, one or the other, because both cannot be achieved. Central planning—ostensibly a method of economic production—is nothing more than a technique of political control.

Our Intourist guide, Lali, a cultivated Georgian, told us on the bus that 20 percent of the houses in Tbilisi are privately owned (presumably they are inherited, not sold in a housing market). She was obviously proud of Georgia, often carefully distinguishing it from Russia.

"I could live here," said an expansive Arnold Beichman one day. "They're nice people and they're obviously not Communists!" He foresaw a great increase in U.S. tourism to the Eastern bloc. (This was about three weeks before the Chernobyl disaster.) "You can have a drink here without worrying about someone dropping some plastique."

We went to Gori, the birthplace of Stalin, thirty-five miles away. On the way we stopped at an ancient monastery perched on a hilltop, and at an eleventh-century Orthodox cathedral. Inside, an old man was kneeling, holding a candle, and crossing himself whenever some East European Party members, who had arrived in a 1950s-looking Chaika limousine, came too close.

Our Gori goal was the Stalin museum. First we saw the log cabin where Stalin grew up. Sometimes it seems there must be a Humble Origin Identikit for World Famous People: bare floorboards (please ignore the velvet rope), simple pitcher on functional table, chair for Papa Stalin. Everything scrubbed, righteous and minimal—somehow fit for the Museum of Modern Art.

A lady with yellowish hair showed us the museum. She greatly admired Stalin, obviously, and was not aware that many of us had come to gawk at the relics in a spirit of ironic amusement. (Here they were honoring the world's greatest dicatator? Without realizing how reprehensible he was!) This may well have been a mistaken attitude on our part, however. Was Stalin different in kind from other Soviet rulers, or only in degree?

Stalin (born 1879) had a beautiful velvet voice, the lady told us. He was a poetic child—a dreamer. His first poems patriotic, people should struggle for a better life. Here were his favorite books from the library. These were members of a Marxist group in Tbilisi; they studied at the seminary with Stalin. Seminary today is an art museum. Stalin became a professional revolutionary at age fifteen; expelled from seminary because of his revolutionary activities there. (In West today, of course, seminary revolution is at the level of doctrine.) Stalin, like Castro and Ortega, was imprisoned and treated all too leniently by the regime he would later help to overthrow.

Beichman asked the guide: Premier Khrushchev's speech denouncing Stalin was in 1956. The Stalin museum opened in 1957. How come?

"We are the people of contradictions," she said.

* * *

The Georgia Museum of Art, Lenin Square, Tbilisi. "Stalin lived here, 1894–99." Here was a fresco of the Archangel Gabriel, Lali proudly told us. There was an icon of the Transfiguration from a nearby monastery. Here the famous Khalkuli triptych—notice the cloisonne enamel technique—incorporating a tenth-century icon of the Virgin. "Georgia was committed to be the country of Our Lady," Lali told us. Now look at this, a sixth-century icon, not painted by human hand according to legend—one of those miraculous icons. . . .

Could she tell us what particular room Stalin lived in, someone asked? No, it was kind of a boarding school, you see.

Dutifully, she took us upstairs to see the modern works. "This is the period of the so-called socialist realism in art," she said. "I don't think you need an explanation." It was the icons that she had wanted to show us.

Back at the hotel, the word was that Senator Kennedy had been on Moscow TV, urging the U.S. to stop nuclear tests. On a tour bus we saw a bumper sticker: San Franciscans for a Nuclear Freeze. And in the dining room we met some homebodies from the American-Soviet Friendship Society. They were on an enraptured three-week, *Nation*-advertised tour of the Soviet Union. Those we spoke to wanted us to know that everything they had seen to date was wonderful: no unemployment, no crime, no profiteering, and don't forget: twenty million killed in World War II. Hearing that I was a journalist a lady from New York who refused to give her name loudly said: "Tell them that unless we live in peace with the socialist countries we shall all die."

By how much should we disarm?

"Fifty percent." This overruled by another voice. "A hundred percent."

Any repression in the Soviet Union?

"No sir!"

Religious persecution?

"No sir!"

Anti-semitism?

"No sir!"

"Oh this is delicious ginger ale," one of their party said. "Boy, do I like to take a drink of ginger ale."

Everything they had seen was wonderful and everything would continue to be wonderful. Upcoming on their itinerary was Kiev. That too would be wonderful. They had already been on Soviet television and no doubt would be again.

Lali, sitting a couple of tables away, had noticed our altercation. I went over to explain what it was that her American charges were arguing about. In the Soviet Union, I said, Communism is compulsory—at least for the ambitious. In America it is voluntary. Quite a few Americans still do have faith in it—for example those we had just been speaking to.

Lali was interested. She seemed a little surprised.

"They don't call themselves Communists," I said. "They don't even think of themselves as Communists. But their state of mind is Communist. They think that a Communist society can be achieved without the use of force."

Lali discreetly said nothing. But she listened attentively, and a look of recognition came across her face when I said that their "state of mind" was Communist. The phrase is Gus Hall's.

I suspect that on any given day the Soviet Union is honeycombed with such American true-believers, whose faith is constantly needed to lift Soviet spirits weighed down by the leaden weight of Communist reality. Chernobyl will have reduced their numbers (and that only temporarily) but not their outlook. Such people are radically hostile to a free society, which they perceive above all as purposeless: failing to insist that all believe the same thing and work collectively to attain it.

In Leningrad we had met an American woman who was attempting to set up an Alcoholics Anonymous chapter, with the approval of the Soviet authorities. Despite this, she had been subjected to a humiliating body search when customs guards saw the Radio Shack computer she had brought with her.

Was not AA based on a belief in a Supreme Being, Larry Moffitt asked her.

Yes, she said, but they were modifying it for Soviet consumption.

"You may be removing the active ingredient that makes it work."

"It's either that or no AA."

She asked Moffitt how he had found the Soviet Union.

"Well, repressive."

"You're just saying the same old thing."

"That's because it's been true for a long time."

Had she taken a look at Soviet housing, Larry asked? Did she know they often had three or four families living in one apartment?

She said that they had poor housing in San Francisco, too.

"Lady, they strip-searched you for bringing in a lap-computer. Doesn't that tell you something."

He was "brainwashed by anti-Soviet propaganda," she rejoined.

He said that she was "the victim of a self-induced blindness."

"And with that," Larry told me, "we reached a stony impasse."

Driving away from the Moscow Airport to our hotel, Olga (who was with us for the entire trip) allowed two more guides onto the bus for the ride into Moscow. One young woman sat next to Beichman. She might quite reasonably have assumed that we were one more peace group.

"Why can't we live in peace?" She pertly asked him.

"Because the Red Army is in places where it isn't wanted."

"Like where?"

"Hungary."

We drove through muddy miles of socialist realism—dreary blocks of flats, Lenin in repose, Lenin making a point to the multitudes, Lenin in profile, Lenin Triumphant. Buses packed with strap-hanging proles were lumbering outward to communal apartments. Everything was dirty, dun-colored, unpainted, grimy, poor, crumbling, trodden down. Old ladies with black shawls and shopping bags plodded along minding their own business. Long Live the Communist Party of the Soviet Union! rang out the building-top slogans. Still not a leaf in sight.

Moscow is a concentric city, and quite abruptly we were in the center, with silvery church domes and freshly gilded onions glinting through bare birch branches. The Kremlin! A high brick wall and a Red Star, a river embankment and here was our hotel, the Rossiya, only two or three hundred yards from St. Basil's Cathedral. It looked, as Christopher Booker has written, "like nothing so much as a cluster of tethered Montgolfier balloons." It was smaller than one had imagined, just as the Kremlin itself was somehow more impressive.

At supper, live music was furnished by a hefty pink-gowned quartet—the Kremlinettes, as I thought of them—playing Moscow rock-cum-pasodoble. They seemed to displease Olga. I think she thought them *niekulturnye*. Holger Jensen of *The Washington Times* gallantly danced with a Russian woman who identified herself as a "commandante from the Urals." Holger speaks Russian.

"Commandante of what?" he asked.

"We have those."

"Are you glad to be in Moscow?" he asked.

"Of course. Lenin is here."

Holger told us that evening that during his earlier stay in the Soviet Union as an AP correspondent, he had managed to wrangle a trip to Siberia. He claimed to have found peasants armed with muskets in the backwoods who were surprised to hear that Russia was no longer ruled by the czars.

The food was not nearly so plentiful as in Tbilisi. In Moscow a second cup of coffee cost you kopecks out of your pocket. But there was always plenty of bread, and (wherever we went) ample butter. The latter is furnished at highly subsidized prices by the European Common Market—mostly West Germany. (The Common Market should be thought of as a system whereby West European taxpayers subsidize both their own farmers and Soviet consumers. "The latest butter shipment to Russia will cost the Community budget $200 million," according to *The Economist* of May 17, 1986.)

Lenin's tomb was not on our tour. Were it not for the two-hour wait in line I would gladly have viewed the mummified founder. I gather there is some doubt as to whether it really is Lenin's corpse. Larry Moffitt said that on an earlier tour he viewed the remains "to make sure he hasn't risen."

After supper, Arnold Beichman and I went out into Red Square. It was cold and the hour late, but still there was a crowd waiting for the goose-stepping guards who change on the hour outside the portal to Lenin's Mausoleum. Death watch duty at the heart of the Evil Empire! The guards seemed so young—still in their teens. People in the crowd spoke to one another, if they spoke at all, in hushed voices, as one might in church. This was Arnold's first time in Red Square, but he had little to say, reacting as though he had seen it all before.

The next day we were granted an audience with the U.S. Ambassador, Arthur Hartman, at his spacious diplomatic residence, Spasso House. First we were served sandwiches and coffee—by Russian maids. Did I not spot the lynx-eyed commandante from the Urals out of the corner of my eye, slipping quietly into a distant pantry? Here was penetration! (But fear not, readers, the State Department knows what it is doing.) Somewhere offstage a piano was being tuned, for the benefit of Maestro Horowitz, shortly to make his long-awaited reappearance in the country of his birth.

His Excellency appeared and spoke to us for twenty minutes or more about the state of U.S.-Soviet relations, and so on, but at the end we were told that his remarks were "off the record," so I cannot divulge his insights, however perspicacious, even if the Russian maids *were* listening in. At one point Hartman's wife came by, needed the car because (Gawd what a bore) she had to go to lunch with the Bulgarians.

Natasha, our attractive Intourist guide (as you see, we had "local" guides, who were always subordinate to Olga, our commandante from Leningrad), took us on a bus tour of Moscow, somewhat spoiled by the overcast, drizzly weather. Like Lali, Natasha knew a lot about the history of art. With politics off-limits, I suspect most non-Party members stop thinking about political affairs and so have plenty of time left over for browsing through art history books and dreaming about the past. One might say that people who believe in Communism dream about the future; those who experience it dream about the past.

Even in steady rain on a Saturday, the streets were packed with people: better than being cooped up in a shared apartment, after all. Dixy Lee Ray commented on the absence of bicycles, as earlier she had noted the absence of any sign of animal transportation in the Georgia countryside.

We drove by the huge Moskva Swimming Pool, site of the Cathedral of Christ the Redeemer, which was demolished by Stalin in 1934. The plan was to put in its place a giant Palace of Soviets, topped by a statue of Lenin twice the height of the Statue of Liberty. But the Great Patriotic War intervened, by which time only the foundations were in place. These were then torn up and used by the military. Today the swimming pool marks the spot. Someone said that when the mists rise up from the water's surface, bathers are secretly baptized.

Natasha told us that there were 40 functioning churches in Moscow, two of which I visited, the St. Louis (Roman Catholic) Church, behind the Lubianka Prison and KGB headquarters, and the Russian Orthodox Yelokhovsky Cathedral, which was a couple of miles from the center of the city. Scaffolding around the dome indicated state maintenance of the Orthodox church. The interior was beautifully kept up. For the 6 pm service there were at least 500 people in the church, mostly, it is true, old ladies or "babushkas" in shawls; hundreds of them were shuffling about, none more than five feet tall, lighting candles, kissing icons and crossing themselves every few minutes. The service lasted for two and a half hours with no congregational singing or participation to speak of. Black-hatted metropolitans ruled the roost as firmly as the Communists do the citizenry.

Occasionally one or another berobed clergyman disappared behind the lateral iconostasis and chanted out from some hidden recess. There was a first-rate choir and a half-hour sermon. There *were* younger people, some conspicuously well dressed, who came and went according to their own schedule. At one point a couple brought in a baby to receive communion (they put a small spoonful of bread and wine into the baby's mouth). A vigilant babushka reprimanded me in sign language for standing with my hands behind my back. There are, incidentally, no pews or chairs in Russian Orthodox churches. My legs grew very tired indeed. Curiously, the service seemed to have no well defined conclusion. It just petered out. While others were still officiating, a clergyman came out from behind a side screen and the crowds pressed close to kiss his hand (no ring on it).

The Kremlin! What a fantastic sight it is. Nothing had prepared me for it. From our hotel we had a panoramic view. Edifices of the Soviet state stand shoulder to shoulder with gold- and silver-domed cathedrals, all meticulously preserved and maintained and gleaming in the pale sun: the Cathedral of the Assumption; the Presidium of the Supreme Soviet; the Cathedral of the Archangel; the Palace of Congresses; the Cathedral of the Annunciation; the House of the Soviets; the Savior Tower; the Statue of Lenin.

Here was the greatest internal contradiction, the ultimate dialectic: dialectical spiritualism! There has to be some great undisclosed meaning to Communism. It is entirely inadequate to think of it as a misguided method of economic production. It is surely something more than that, and something more sinister. I walked the mile and a half around the Kremlin wall one afternoon. Strange thoughts come to you. Perhaps it helps not to be distracted for a while by our own endless media barrage of half truth. Keeping me company was the endlessly patient crowd of Russians trudging along, lost in their own thoughts. There's no penalty for thinking, and Vladimir Bukovsky for one has said that among people who trust one another the conversation in Moscow is the world's best.

In the pre-revolutionary phase of Russian foreboding about what was to come, Dostoevsky among others saw a connection between the decline of Christianity and the rise of "future socialism," which was destined to arise "in lieu of the deteriorated Christian principle." At the same time he believed that "God-bearing" Russians were destined to "regenerate and save mankind in the name of a new God."

The thought arose, as one walked along with these poor Russians: Could it be that they, having for seventy years experienced the ordeal of Communism, have by their suffering helped to "regenerate" mankind?

Was this indeed the meaning of the fifteenth-century prediction, made after the fall of Constantinople, that Moscow would be the "third Rome"?

More recently the Russian writer Nikolai Berdyaev has made the point that Communism is a form of chastisement. "Christian justice has not worked itself out fully in life," he wrote in 1931, "and in virtue of the mysterious ways of Divine Providence the forces of evil have undertaken the task of realizing social justice." For this reason, he said, Communism "should have a very special significance for Christians, for it is a reminder and a denouncement of an unfulfilled duty."

My conviction under the Kremlin wall was that Communism was destined to spread further and would engulf more nations. Certainly in the U.S today we have no principle of resistance to it. Communism cannot be stopped by tax cuts, as the supply-siders believe, nor can it be assuaged by negotiated settlements, as the liberals hope. If Berdyaev is right, not to mention Whittaker Chambers and others, Communism is an evil religion, the "reciprocal" of Christianity, which in the event of Christian default plays a mysterious but necessary role in the divine plan. Looked at this way, Communism can only be resisted by a restoration of Christian principle and practice in the West.

My second Kremlin conviction, shared by other visitors to the Soviet Union, notably Malcolm Muggeridge, was that Christianity will one day be officially restored in Russia. Russia will be converted, as the visionaries at Fatima, Portugal, reported in 1917. It is not at all a modern thought, that the Russian Orthodox churches now being restored (ostensibly to give the tourists something to see) are subconsciously being readied for a return to their original use. Nor is it necessarily for us an encouraging thought, because so dramatic an event is only likely to occur at a moment of dire national emergency—for example when the Western countries that have all along been subsidizing Soviet Communism are no longer able to do so.

Postscript

All of Larry Borenstein's predictions about the value of real estate investment in the French Quarter turned out to be true. Inflation was just over the horizon when I talked to him in 1973. His portfolio of buildings must have been worth millions by the time Ronald Reagan was elected in 1980. But Larry never stopped working long enough to enjoy his fortune. He had first one heart attack, then another, and still he didn't slow up or stop smoking.

According to Allan Jaffe, his friend and sometime business associate, on January 6, 1981, Larry was in his house on Royal Street and being interviewed by a young woman (for what publication or research project is not clear), when he died suddenly of a heart attack. He was 61 years old.

Allan Jaffe, the owner and manager of Preservation Hall in New Orleans, was himself diagnosed with cancer in December 1986. He died less than three months later, at the age of 51. Then, in June 1987, the old-time trumpeter Kid Thomas Valentine died in New Orleans, at the age of 91. Kid Thomas was perhaps the first jazz musician that Borenstein had hired to play when Preservation Hall was known as Associated Artists Studio in the late 1950s. One is tempted to say that with the death of Kid Thomas an era really did come to an end. but only last week I went to hear the touring Preservation Hall Band at Stanford University's Laurence Frost amphitheater for their usual Fourth of July concert, and they sounded fine.

My old friend, the artist Noel Rockmore, still lives in the French Quarter and he still does occasional portraits and pastel sketches of Bill Russell, who is now 82 years old. When I called, Rockmore told me that Russell was on tour in Europe, playing violin with the New Orleans Ragtime Orchestra. Rockmore said that Bill was still "essentially the same, perhaps a little more frail." But others I spoke to said that Bill's asthma and excema are now a good deal worse, and that he is much weaker. He still spends most of his time in his apartment on Orleans Street, in the French Quarter, surrounded by all his papers and boxes. People try to tell him he shouldn't be in there with all that dust. He says it's all right as long as his boxes don't get moved about too much.

Jim Garrison was elected judge on Louisiana's Fourth Circuit Court of Appeals in 1978, having been defeated for reelection as district attorney in 1973. He is 65 years old and divorced from his second wife. His five children are all in college. "I'm a bachelor," he told a reporter from the New Orleans *Times Picayune* in 1986, "and I'd like to remain an eligible bachelor. Why don't you just say I look a lot younger than I am?" Garrison remains convinced that Clay Shaw (who died of cancer in 1973) was guilty of conspiring to assassinate President Kennedy, and that Lee Harvey Oswald was framed.

"There's a possibility I was wrong," he told Elizabeth Mullener of the *Picayune*. "May I describe that possibility? That possibility is exactly equal to the possibility that the sun will rise in the west tomorrow morning. Now you wake up early and see where the sun rises tomorrow morning, and if it rises in the West then I'm wrong." Mullener reported that Garrison "builds up a formidable head of steam" when he talks about the Shaw case, "firing off names and dates and places effortlessly. His holistic theory links the CIA not just with the Kennedy assassination but with a procession of world events that preceded and followed it." Many other books about the assassination were scattered about his office. Garrison referred to them as "disinformation projects" generated by the CIA to cover its tracks.

The Bail Reform Act, whose bad effects I encountered in the District of Columbia, was itself reformed in 1984. The new law permitted the detention of defendants charged with serious crimes, provided the prosectors could show that "the safety of any other person and the community" could not be protected if the defendant was freed. The U.S. Court of Appeals for the Second Circuit (Manhattan) then struck down the 1984 law as unconstitutional, on Fifth Amendment grounds. But on May 27, 1987, the U.S. Supreme Court reversed the appeals court by a 6–3 vote. Criminal defendants considered a threat to public safety may now be jailed before trial. *The New York Times* called it "one of the most important criminal law decisions in years."

* * *

I don't know what Sonia Johnson is up to today. She received 72,200 votes in the 1984 election—0.08 percent of the vote. This was almost exactly double the vote received by Gus Hall, the Communist Party candidate, but only one-third the number received by the 1980 Citizens Party candidate, Barry Commoner. In percentage terms, Sonia Johnson's Citizens Party did best in Louisiana (0.6 percent of the vote), and also did well in California, where she received 26,287 votes. As far as *The New York Times* was concerned, Sonia was last heard from filing suit in Federal Court, charging that she had been unconstitutionally excluded from the presidential debates. I'm sure it won't be long before we hear from her again.

Eric Hoffer died in his San Francisco apartment in April 1983. He had been in a general decline and had spent some time at the Children's Hospital, but evidently he didn't like it there or he preferred to be at home. Lili Osborne was beginning to spend a good deal of time looking after Eric. Selden Osborne was helping out, too. Eventually Lili decided that she would have to move in a cot so that she could spend the night. Selden was alone with Eric in the apartment when Lili was gone. Eric was not in very good shape. At one point he said that he wanted to go to the bathroom and Selden (by then in his early 70s) worried how he would lift Eric if he should fall. But Eric made it back to his bed safely. Selden heard him breathing as they lay in the dark. Later Selden woke up and Eric's breathing could be heard no more. He was gone—you could say that he didn't say goodbye to anyone.

In April 1987, I drove around San Francisco with Selden Osborne. He showed me places on the waterfront where he and Eric used to work. Longshoremen no longer work there. They have nearly all migrated to Oakland. Selden told me that in his time an important union activity, and one that he now questions, was impeding the introduction of new technology—container cargo in particular. Selden is in good health and still as radical as ever, going on peace marches and sometimes corresponding with Daniel Ellsberg. Stopping the Strategic Defense Initiative is now his great and overriding cause.

I also spent some time with Lili Osborne, her brother and sisters—the Fabilli family. One day I asked Lili if Eric had believed in God. Eric's notebooks, only small portions of which have been published, have recently been typed up, and in response to my question Lili a few days later sent me an excerpt from Notebook 91, August 7, 1966 to November 25, 1966:

"I don't have to believe in him. I live with him. Jehovah, man's most sublime invention, is my God. He is with me in all my thought. And I treat him as a God and not an errand boy to fetch what I need or nurse me or defend me or comfort me, No! He and I discourse about the wonder that is man, that I am in him and he in me."

Lili saw to it that Eric had a Catholic funeral. He was buried in Holy Cross Cemetery.

David Brodsly, who wrote *L. A. Freeway*, is still working for the city of Los Angeles. He is a principal financial analyst for new city projects. He told me that the Century Freeway, a 17-mile stretch from Norwalk to the airport, is now under construction. It is "far and away the most expensive freeway" the city will have built, he said. The city has also broken ground for the first four miles of metro-rail. I told Brodsly that his observation about the city planners failing to foresee the freeway system in the 1940s, thus allowing it to be started (if its full extent had been foreseen, it would have seemed prohibitively expensive to legislators), had been passed on to those pressing President Reagan for immediate deployment of the Strategic Defense Initiative. There exactly the same lesson applies. Waiting for massive projects to be fully planned effectively kills them.

I looked forward to seeing Francis Coppola's film *One from the Heart*, but it was gone from the movie theaters before I had the chance. According to a magazine article by Michael Covino, published in 1987, the film "proved just about unreleasable. The slight romantic comedy plays second fiddle to the electronics extravaganza—the camera steals the show, zooming and doing elaborate tracks all over the place. After running up costs of $27 million, *One from the Heart* took in about a million at the box office." Pauline Kael wrote in *The New Yorker*: "The movie isn't from the heart, or from the head, either: It's from the lab."

I'd still like to see it. And by the way, the soundtrack does seem to have improved in recent films I have seen.

Simon Rodia's Watts Towers still stand and seem to be flourishing. I saw an article in *The Los Angeles Times* about some art-crowd cultural event held at the towers a few months ago.

Mitch Snyder has gone from strength to strength. In 1986 he began fasting again, whereupon the Reagan Administration caved in twice, giving him everything he wanted: first the dilapidated building occupied by the homeless, then $5 million to repair it. At the end of June 1987, I phoned Snyder at the shelter. He told me he felt "old, hot and tired." The White House officials he had dealt with when fasting had been Alfred Kingon and Dennis Thomas, but he was "certain" that White House chief of staff Donald Regan had been the one giving the orders to yield. (Surrender came only four days into Mitch's second 1986 fast.)

In my earlier conversations with Mitch, he said he envisioned a voluntary program, with the government providing buildings only. But in February 1987, President Reagan signed emergency legislation transferring $50 million from disaster relief to the homeless. By early March a number of Congressmen, including Joseph P. Kennedy II, were spending the night on grates in the nation's capital. Later that month, first the House of Representatives and then the U.S. Senate authorized over $400 million for aid to the homeless. Mitch Snyder was confident (with ample precedent) that Reagan would sign the legislation.

"When you consider that the city of New York is spending $240 million," Snyder told me, "$450 million nationally is not exactly a large sum of money. It simply represents a start—a reluctant willingness to get involved."

Now that he had so successfully enlisted the government (and taxpayers' money) on his side, I wondered if he would be looking for new fields to conquer. He replied that all the issues that interested him were really one. "The ills that befall our nation stem from the same values and priorities," Mitch said, more in sorrow than in anger. "They're all essentially the same, whether its homelessness or nuclear proliferation, violation of the environment or gross inequities in the distribution of wealth of our land and our Earth. They're all manifestations of a common denominator. What we're doing is not separable from those fighting nuclear power or nuclear weaponry."

About a year after I had interviewed him, I called Tom Ryan at the home for AIDS patients run by Mother Teresa's nuns. He said he had recently been feeling weaker, but that day he seemed better. "I'm the only one left from when you were here," he told me. All the others had died, including Walter and Juan. He said that most of those who came to the house convert to Catholicism. "There's a confirmation coming up on Friday for two of the guys," he said. Tom was looking forward to a visit from Mother Teresa in August.

Sister Dolores has moved to Washington, D.C., where the Missionaries of Charity have opened another house for AIDS patients. And now comes word that Archbishop John Quinn has asked Mother Teresa to open a similar house in San Francisco. "Turn to God," was Tom's parting advice. "There is no other way."

In my *American Spectator* article on the Soviet Union I commented on "the lynx-eyed commandante from the Urals, slipping quietly into a distant pantry." Six months later, of course, quite a few details about the Soviet espionage ring in the U.S. Embassy in Moscow were made public. A bevy of *femmes fatales* was unmasked, U. S. Marines were charged, cover stories were rushed into print in the news magazines. How did I manage to scoop the field on this important story? Perhaps I had better not say. But I can divulge here that I will be returning to England for a nostalgic visit in a few weeks, this time by jet (the old *Queen Elizabeth* was destroyed by fire in Singapore), and probably not first class. Henry Pleasants, the music critic and diplomat, took up residence in London not long after I met him at the Wigmore Hall, and he lives there to this day. I do hope that we will be able to have a reunion. There should be a lot to talk about, after 25 years. . . .

July 1987